ANDREW GRAY

A TREATISE ON SPINNING MACHINERY

ILLUSTRATED WITH

PLANS OF DIFFERENT MACHINES MADE USE OF IN THAT ART

Elibron Classics
www.elibron.com

Elibron Classics series.

ISBN 1-4021-5927-7 (paperback)
ISBN 1-4021-0401-4 (hardcover)

This Elibron Classics Replica Edition is an unabridged facsimile of the edition published in 1819 by Archibald Constable and Co.; Longman an
Edinburg; London.

JANUARY, 1819.

MODERN PUBLICATIONS,

AND

NEW EDITIONS OF VALUABLE STANDARD WORKS,

PRINTED FOR

LONGMAN, HURST, REES, ORME, AND BROWN, LONDON.

Messrs. LONGMAN, HURST, REES, ORME, *and* BROWN, *having just completed and fitted up elegant and commodious Retail Rooms, retired from their Wholesale Business, for the purpose of exhibiting their extensive and valuable Stock of Books, including an unprecedented Collection of the Rarities and Curiosities of Literature, they take the Liberty respectfully to invite their Friends, and the Amateurs of Literature, to view these interesting Literary Treasures.*

A SECOND JOURNEY THROUGH PERSIA TO CONSTANTINOPLE,

Between the Years 1810 and 1816. With a Journal of the Voyage by the BRAZILS and BOMBAY to the PERSIAN GULPH; together with an Account of the Proceedings of his Majesty's Embassy, under his Excellency Sir Gore Ousley, Bart. K.S.L.
By JAMES MORIER, Esq.
Late his Majesty's Secretary of Embassy, and Minister Plenipotentiary to the Court of Persia.
In Royal 4to. with Maps, coloured Costumes, and other Engravings from the Designs of the Author, Price 3*l.* 13*s.* 6*d.* Bds.

A JOURNEY FROM INDIA TO ENGLAND,

Through Persia, Georgia, Russia, Poland, and Prussia, in the Year 1817.
By LIEUT.-COL. JOHNSON, C. B.
In 4to. illustrated with numerous Engravings, Price 2*l.* 2*s.* Boards.

MEMOIRS OF THE COURT OF QUEEN ELIZABETH.

BY LUCY AIKIN.
In 2 Vols. 8vo. 1*l.* 5*s.* Bds. with a Portrait. 3d Edit.
The present Work is composed upon the plan of uniting with the personal History of a celebrated female Sovereign, and a connected Narration of the domestic Events of her Reign, a large Portion of Biographical Anecdote, private Memoir, and Traits illustrative of the Manners, Modes of Thinking, and Literature of an interesting Period of English History.

A NARRATIVE OF SHIPWRECK OF THE OSWEGO,

On the Coast of South Barbary, and of the Sufferings of the Master and the Crew while in bondage among the Arabs; interspersed with numerous Remarks upon the Country and its Inhabitants, and the peculiar Perils of that Coast.—By JUDAH PADDOCK, her late Master.
In 4to. price 1*l.* 5*s.*

THE TRAVELS OF MARCO POLO,

A Venetian,
IN THE THIRTEENTH CENTURY;
Being a Description of that early Traveller, of remarkable Places and Things, in the Eastern Parts of the World. Translated from the Italian, with Notes.
By WILLIAM MARSDEN, F.R.S. &c.
In 4to. with a Map, Price 2*l.* 12*s.* 6*d.* or on Royal Paper, 4*l.* 4*s.* Bds.

The Poetical Remains

OF THE LATE
DR. JOHN LEYDEN,
WITH MEMOIRS OF HIS LIFE.
By the REV. JOHN MORTON.
In 1 Vol. 8vo.

TRANSACTIONS OF THE LITERARY SOCIETY OF BOMBAY,

WITH ENGRAVINGS.

NARRATIVE OF A JOURNEY IN THE INTERIOR OF CHINA,

And of a Voyage to and from that Country, in the Years 1816 *and* 1817;
Containing an Account of the most interesting Transactions of Lord Amherst's Embassy to the Court of Pekin, and Observations on the Countries which it visited. By CLARKE ABEL, F.L.S.
And Member of the Geological Society, Chief Medical Officer and Naturalist to the Embassy. In 4to. illustrated by Maps and other Engravings. Price 3*l.* 3*s.* Bds.

THE FAMILY SHAKSPEARE;

In which nothing is added to the original Text: but those Words and Expressions are omitted which cannot with Propriety be read aloud in a Family.
By THOMAS BOWDLER, Esq. F.R.S. and S.A.
In 10 Vols. Royal 18mo. Price 3*l.* 3*s.* Bds.

"My great objects in this undertaking are to remove from the writings of Shakspeare, some defects which diminish their value, and, at the same time, to present to the public an edition of his plays, which the parent, the guardian, and the instructor of youth may place without fear in the hands of the pupil; and from which the pupil may derive instruction as well as pleasure; may improve his moral principles, while he refines his taste; and without incurring the danger of being hurt with any indelicacy of expression, may learn in the fate of Macbeth, that even a kingdom is dearly purchased, if virtue be the price of acquisition."—*Preface.*

AN ACCOUNT OF THE KINGDOM OF CAUBUL,

And its Dependencies
IN PERSIA, TARTARY, AND INDIA;
Comprising a View of the Afghaun Nation, and a History of the Dooraunee Monarchy.
By the Hon. MOUNTSTUART ELPHINSTONE,
Of the Hon. East India Company's Service; Resident at the Court of Poona; and late Envoy to the King of Caubul.
The 2d Edit. in 2 Vols. 8vo. illustrated by Maps and Plates. Price 2*l.* 2*s.* Bds.
A few Copies may be had in 4to. Price 3l. 13s. 6d. Bds.

THE ANNUAL BIOGRAPHY AND OBITUARY, FOR 1819,

With Silhouette Portraits.
Comprehending the Private Life of Her late Majesty, with an historical Dissertation on the Family of Mecklenburgh Strelitz; a Memoir of Sir Samuel Romilly, illustrated by authentic Notes concerning his Family; a Biographical Account of Lord Ellenborough, late Lord Chief Justice of the King's Bench; another of the late Mr. Dempster, with some Original Letters to a Member of His Majesty's Privy Council; a Life of, and Analysis of the Impeachment of Mr. Hastings; Biographical Notices of Dr. Burney, Sir Thomas Bernard, Sir Richard Croft, Mr. Rose, Dr. Cogan, founder of the Royal Humane Society; Dr. Adams, Rev. William Beloe, &c. with an Analytical Account of their Works.
The whole is interspersed with a variety of original documents, together with an account of recent biographical works, and an Index of persons lately deceased,
Also may be had,
Vols. I. and II. for 1817 and 1818, price 15*s.* each.

PATHOLOGICAL AND SURGICAL OBSERVATIONS ON

DISEASES OF THE JOINTS.

By B. C. BRODIE, F.R.S
Assistant Surgeon to St. George's Hospital, and Lecturer on Surgery.
In 1 Vol. 8vo. illustrated by Plates. Price 16*s*. Bds.

A SUCCINCT ACCOUNT of the **CONTAGIOUS FEVER** of this Country, as exemplified in the Epidemic now prevailing in London, with the appropriate Method of Treatment as practised in the House of Recovery.

To which are added.

OBSERVATIONS on the NATURE and PROPERTIES of CONTAGION, tending to correct the Popular Notions on this Subject, and pointing out the Means of Prevention.

By THOMAS BATEMAN, M.D. F.L.S.
Physician to the Public Dispensary, and Consulting Physician to the Fever Institution in London, &c. &c.
Published in 8vo. Price 6*s*.

THE LONDON DISPENSATORY;

Containing—1. Pharmacy—2. The Botanical Description, Natural History, Chemical Analysis, and Medicinal Properties, of the Substances of the Materia Medica—3. The Pharmaceutical Preparations and Compositions of the Pharmacopœias of the London, Edinburgh, and Dublin Colleges of Physicians. The whole forming a practical Synopsis of Materia Medica, Pharmacy, and Therapeutics: illustrated with many useful Tables and Copper-plates of Pharmaceutical Apparatus. By ANTHONY TODD THOMSON, F.L.S. &c. In One large Volume 8vo. (revised and altered according to the last Edition of the London and Edinburgh Pharmacopœias), Price 15*s*. Boards. Second Edition.

*** This edition contains the synonyma of the names of the articles, in the French, German, Italian, Spanish, and East Indian languages.

SPANISH AMERICA;

Or a Descriptive, Historical, and Geographical Account of the Dominions of Spain, in the Western Hemisphere, Continental and Insular; illustrated by a Map of Spanish North America, and the West India Islands; a Map of Spanish South America, and an Engraving, representing the comparative Altitudes of the Mountains in those Regions. By R. H. BONNYCASTLE, Captain in the Corps of Royal Engineers. In 2 Vols. 8vo. Price 1*l*. 1*s*. Bds.

TRAVELS

IN CANADA, AND THE UNITED STATES OF AMERICA,
In 1816 and 1817.
By F. HALL, Esq. late Military Secretary to General Wilson, Governor in Canada.
In 8vo. Price 14*s*. Boards.

AN AUTUMN NEAR THE RHINE;

Or, Sketches of Courts, Society, and Scenery, in some of the German States bordering on the Rhine.
In 8vo. Price 14*s*.

LETTERS

FROM THE ABBE EDGEWORTH TO HIS FRIENDS, Written between the Years 1777 and 1807; with Memoirs of his Life, including some Account of the late Roman Catholic Bishop of Cork, Dr. Moylan, and Letters to him from the Right Hon. Edmund Burke, and other Persons of Distinction.
By the Rev. THOMAS R. ENGLAND.
In 8vo. Price 8*s*. Boards.

TRAVELS

THROUGH

SOME PARTS OF GERMANY, POLAND, MOLDAVIA, AND TURKEY.

By ADAM NEALE, M.D.
Late Physician to the British Embassy at Constantinople, and Physician to the Forces.
In 1 Vol. 4to. illustrated by 11 Plates. Price 2*l*. 2*s*. Bds.

THE FAST OF ST. MAGDALEN.

A Romance.
By MISS ANNA MARIA PORTER.
The 2d Edition. Three Vols. 12mo. Price 1*l*. 1*s*. Bds.

MEMOIRS of the late Mrs. **ELIZABETH HAMILTON**; with a Selection from her Correspondence, and other unpublished Writings.
By MISS BENGER.
2d Edit. In 2 Vols. 8vo. with a Portrait after Raebairn, 1*l*. 1*s*. Bds.

NEW TALES.

By Mrs. OPIE. 3d Edition. In 4 Vols. 12mo. Price 1*l*. 8*s*. Bds.

Contents.—Mrs. Arlington; or, all is not Gold that glitters—Proposals of Marriage—White Lies—Henry Woodville—The Young Man of the World—A Tale of Trials—An Odd-tempered Man—The Ruffian Boy, a Tale founded on fact—The Welcome Home; or, the Ball.

SOPHIA;

OR, THE DANGEROUS INDISCRETION.
A Novel. In 3 Vols. 12mo. Price 16*s*. 6*d*. Bds.

THE FUDGE FAMILY IN PARIS,

IN A SERIES OF LETTERS,
FROM
Phil. Fudge, Esq.—Miss Biddy Fudge—Mr. Bob Fudge, &c.
Edited by THOMAS BROWN, the Younger, Author of the Two-Penny Post Bag.
In Foolscap 8vo. 9th Edit. Price 7*s*. 6*d*. Boards.

LALLA ROOKH;

An Oriental Romance.
By THOMAS MOORE, Esq.
8th Edition, in 8vo. Price 14*s*.
Also just published, in 8vo. Price 12s.
ILLUSTRATIONS of the POEM, engraved by CHARLES HEATH, from Paintings by R. WESTALL, R. A.
A few Proofs, in 4to. may be had, Price 1*l*. 5*s*. and the Poem, 2*l*. 2*s*. Bds.

LIONEL;

OR, THE LAST OF THE PEVENSEYS.
A Novel. In 3 Vols. 12mo. Price 1*l*. 1*s*. Bds.

THE

BACHELOR AND THE MARRIED MAN;

OR,

THE EQUILIBRIUM OF THE "BALANCE OF COMFORT."

A Novel. 2d Edit. In 3 Vols. 12mo. Price 16*s*. 6*d*. Bds.

THE PHYSIOGNOMIST.

A Novel. By the Author of "The Bachelor and the Married Man." In 3 Vols. 12mo. Price 16*s*. 6*d*. Bds.
In the Press, by the same Author,
"HESITATION, or, TO MARRY OR NOT TO MARRY."

CORRECTION. A NOVEL.

2d Edit. In Three Volumes, 12mo. Price 1*l*. 1*s*. Bds.

"The seducing images of luxury, of splendour, and of homages, of power and independence, are too seldom counteracted by the only preservative—a religious education. The world is too generally entered upon as a scene of pleasure instead of trial; as a theatre of amusement, not action."—*Hannah More.*

A NEW NOVEL is preparing for the Press by the same Author.

THE PERSONAL NARRATIVE

OF M. DE HUMBOLDT'S TRAVELS TO THE EQUINOCTIAL REGIONS OF THE NEW CONTINENT,
During the Years 1799—1804.
TRANSLATED BY HELEN MARIA WILLIAMS,
Under the immediate Inspection of the Author.
Vols. I. II. & III. in 8vo. with Maps and Plates.
Price 1*l*. 19*s*.
Volume IV. it is expected, will appear in January.
By the same Author,

The RESEARCHES, comprising the Text of the Atlas Pittoresque, and a Selection of the Plates. 2 Vols. 1*l*. 11*s*. 6*d*.

A POLITICAL ESSAY on the KINGDOM of NEW SPAIN, with Maps, in 4 Vols. 8vo. Price 3*l*. 13*s*. 6*d*. Bds.

A DESCRIPTION

Of the Character, Manners, and Customs of the PEOPLE OF INDIA;

And their Institutions, Religious and Civil.
By the ABBE J. A. DUBOIS, Missionary in the Mysore.
Translated from the French Manuscript.
In 4to. Price 2*l*. 2*s*. Boards.

A GENUINE AND CORRECTED REPORT
OF THE
SPEECHES
OF THE
LATE RIGHT HONOURABLE WILLIAM PITT,
In the House of Commons. In 3 Vols. 8vo. Price 1*l.* 16*s.* Boards, the 3d Edit.

THE SPEECHES
OF THE
RIGHT HON. CHARLES JAMES FOX.
In the House of Commons. In 6 Vols. 8vo. with an Introduction by the Right Hon. Lord Erskine, 4*l.* 4*s.* Bds.

THE SPEECHES
OF THE
RIGHT HONOURABLE EDMUND BURKE,
In the House of Commons and in Westminster Hall. In 4 Vols. 8vo. Price 2*l.* 16*s.* in Bds.

SPEECHES IN PARLIAMENT
OF THE
RIGHT HONOURABLE WILLIAM WYNDHAM;
With some Account of his Life.
By THOMAS AMYOT, Esq.
In 3 Vols. 8vo. 1*l.* 16*s.* Boards.

SPEECHES
OF THE
RIGHT HON. JOHN PHILPOT CURRAN,
Late Master of the Rolls in Ireland, on interesting State Trials, &c. To which is prefixed, a Memoir of his Life. In 8vo. (new Edition with Additions) 12*s.* Bds.

THE SPEECHES
OF
CHARLES PHILLIPS, ESQ.
Delivered at the Bar, and on various public Occasions, in Ireland and England. In 1 Vol. 8vo. 7*s.* Bds.

A PRACTICAL DICTIONARY
OF
Popular Medicine,
Comprehending the different Branches of the Healing Art, so far as they relate to the Preservation of the Health of Man, &c.
By RICHARD REECE, M.D.
A new Edit. in 8vo. with Additions. Price 16*s.* Bds.

THE MEDICAL GUIDE,
For the Use of Families and Young Practitioners, or Students in Medicine and Surgery; being a complete System of modern and domestic Medicine; to which are added a Family Dispensatory, and a Copious Appendix; containing explicit Instructions for the ordinary Management of Children, and such Cases or Accidents which require immediate Aid, &c.
By RICHARD REECE, M.D.
Fellow of the Royal College of Surgeons.
The Eleventh Edition. In 8vo. Price 10*s.* 6*d.* Bds.

Delineations
OF
THE CUTANEOUS DISEASES,
Comprised in the Classification of the late Dr. Willan: including the greater Part of the Engravings of that Author, in an improved State, and completing the Series as intended to have been finished by him.
By T. BATEMAN, M.D. F.L.S.
Physician to the Public Dispensary, and to the London House of Recovery.
In One Vol. 4to. with upwards of 70 coloured Plates, Price 12*l.* 12*s.* Bds.
The Series of New Engravings, representing those Diseases which should have been figured in the subsequent Parts of Dr. Willan's unfinished Work, may be had by the Possessors of that Work, separate, 7*l.* Bds.

SYSTEMATIC EDUCATION;
Or, Elementary Instruction
In the various Departments of Literature and Science, with Practical Rules for studying each Branch of useful Knowledge.
By the Rev. W. SHEPHERD, the Rev. J. JOYCE, and the Rev. LANT CARPENTER, LL. D.
In Two thick Vols. 8vo. (closely printed, and illustrated with Nine Plates, by LOWRY, &c.) The Second Edit. improved. Price 1*l.* 11*s.* 6*d.* Bds.

AN ESSAY
ON
THE CHEMICAL HISTORY AND MEDICAL TREATMENT OF
URINARY CALCULI.
By ALEXANDER MARCET, M.D. F.R.S.
One of the Physicians to Guy's Hospital.
In Royal 8vo. with 10 Plates, comprising upwards of 20 Figures, many of which are beautifully coloured, Price 18*s.* Bds.

SERMONS.
By the late Rev. WALTER BLAKE KIRWAN,
Dean of Killala.
WITH A SKETCH OF HIS LIFE.
In 1 Vol. 8vo. With a Portrait of the Author. 12*s.* Bds.

THE
HISTORY OF ALL RELIGIONS,
With Explanations of the Doctrines and Order of Worship, as held and practised by all the Denominations of professing Christians.
By JOHN BELLAMY.
The Second Edition, with considerable Alterations and Additions, in 12mo. Price 6*s.*; and in 8vo. fine Paper, and hot-pressed. Price 10*s.* 6*d.* Bds.

RURAL PHILOSOPHY;
OR,
Reflections on Knowledge, Virtue, and Happiness;
Chiefly in Reference to a Life of Retirement in the Country. By ELY BATES, Esq.
The 6th Edition. In 8vo. Price 9*s.* in Bds.

SCRIPTURAL ESSAYS,
Adapted to the Holidays of the Church of England: with Meditations on the prescribed Services.
By Mrs. WEST, Author of Letters to a Young Man, &c.
In 2 Vols. 12mo. Price 12*s.* Bds.

LETTERS ON ENGLISH HISTORY,
From the Invasion of Julius Cæsar to the Battle of Waterloo;
FOR THE USE OF SCHOOLS.
By J. BIGLAND. In 12mo. Price 6*s.* Bds.

LETTERS ON FRENCH HISTORY,
From the Earliest Period to the Battle of Waterloo, and re-establishment of the House of Bourbon;
FOR THE USE OF SCHOOLS.
By J. BIGLAND. In 12mo. Price 6*s.* Bds.

COMPENDIUM FLORÆ BRITANNICÆ.
Auctore
JACOBO EDVARDO SMITH, EQU. AUR. M.D.
Societatis Linnæanæ Præside, &c. &c.
The Second Edition, corrected, and continued to the End of the Third Volume of the FLORA BRITANNICA, with all new-discovered Plants from the ENGLISH BOTANY, and References to that Work throughout.
In 12mo. Price 7*s.* 6*d.* Bds.

RODERICK,
THE LAST OF THE GOTHS,
A Tragic Poem.
By ROBERT SOUTHEY, Esq.
Poet Laureate, Member of the Royal Spanish Academy, of the Royal Spanish Academy of History, and of the Royal Institute of the Netherlands.
5th Edition. In 2 Vols. foolscap 8vo. Price 16*s.* Bds.

THE DRAGON KNIGHT;
A Poem, in Twelve Cantos.
BY SIR JAMES BLAND BURGESS, BART.
Elegantly printed in 1 Vol. 8vo. Price 12*s.* Boards.

An INTRODUCTION to ENTOMOLOGY;
Or, Elements of the Natural History of Insects.
Illustrated by coloured Plates.
By WILLIAM KIRBY, M. A. F. R. S. F. L. S. and WILLIAM SPENCE, Esq. F. L. S.
Second Volume. A new Edition. Price 18*s.*
Also, the Third Edition, enlarged and improved, of the First Volume of the above Work. Price 18*s.*
An APPENDIX to the First Edition of Vol. I. comprising the additional Matter inserted in the Second Edition, Price 1*s.* 6*d.*

NEW BOOKS

ON

Agriculture and Rural Affairs,

PUBLISHED BY J. HARDING,

36, ST. JAMES'S STREET.

NEW THEORY OF AGRICULTURE.

A NEW THEORY OF AGRICULTURE, in which the Nature of Soils, Crops, and Manures is explained, many prevailing prejudices are exploded, and the Application of Bones, Gypsum, Lime, Chalk, &c. determined on Scientific Principles. Dedicated to T. W. Coke, Esq. M. P. By WILLIAM GRISENTHWAITE. 12mo. 5s.

Mr. CURWEN'S REPORT.

A REPORT made to the Workington Agricultural Society by the President, J. C. Curwen, Esq. M. P. 8vo. 5s.

This Work gives a detail of the Proceedings at the Schoose Experimental Farm for the last five years, and includes a general view of recent Discoveries and Improvements in Agriculture.

BLAIKIE'S AGRICULTURAL TRACTS.

A TREATISE on MILDEW, and the CULTIVATION of WHEAT, including Hints on the Use of Lime, Chalk, Marl, Clay, Gypsum, &c. By FRANCIS BLAIKIE, Steward and Agent to T. W. Coke, Esq. price 1s. 6d.

An ESSAY on the CONVERSION of ARABLE LAND into PASTURE, by transplanting turf, and on other Rural Subjects. New Edition, by the same, 1s. 6d.

An ESSAY on the MANAGEMENT of FARM-YARD MANURE, and Formation of Compost, accompanied with a Platé, and Description of the Inverted Horse Hoe, invented by the Author. New Edition, 2s.

A TREATISE on the MANAGEMENT of HEDGES and HEDGE-ROW TIMBER. Second Edition, 2s.

POTATOES.

A PRACTICAL TREATISE on the CULTURE and MANAGEMENT of POTATOES, with a Catalogue of the best Kinds, and Hints on the Use of Potatoes in Feeding Stock, &c. 2s.

DAIRY MANAGEMENT.

ESSAYS on MANAGEMENT of the DAIRY, shewing the modern Practice of the best districts in the Manufacture of Butter and Cheese, deduced from a series of observations made during thirty years Practice. By J. TWAMLEY, and others. A New and enlarged edition, with cuts, 7s.

The DAIRY BOOK, or Account of the Produce and Consumption of Milk, Cream, and Butter, 6s.

THE WORKMAN'S ACCOUNT BOOK, ruled for keeping a regular account of labour and wages. Printed on writing paper. 12s. for a Year's Book.

FARMING ACCOUNTS.

HARDING'S ORIGINAL FARMER'S ACCOUNT

BOOK for 1820-21. *A new Edition improved,*

This Work is a collection of Forms ruled for keeping the Accounts of a Country Establishment of magnitude, and being founded upon long experience in Rural Management, will prove highly useful to Landholders, Farmers, Stewards, &c.

Price 21s. for a Year, or 10s. 6d. for a Half Year's Book. Continued annually.

The Work contains Ruled Tables for keeping

A Journal of Occurrences on the Farm, Daily Account of Labour performed, Account of Receipts and Expenses, Weekly Account of Live Stock,	Produce and Consumption of Corn, Hay, Potatoes, and other Crops, Dairy Account, Annual Receipts and Expenses, and sundry other matters.

This Work has now been before the Agricultural Public for many Years, and the plan of it pretty extensively known and approved by Farmers in every part of the Kingdom. A new Edition with improvements is now published, and additions will be made from time to time as required. Gentlemen wishing to possess the original Work should be careful to ask for ***Harding's*** Account Book.

HARDING's FARMER'S MEMORANDUM BOOK

for 1820-21; or, Journal of Country Business and Accounts throughout the Year; consisting of Tables arranged in the most perspicuous and easy manner; this is an abridgement of the above, and is adapted to a Farm of moderate extent, 4to. Price, for a Year, 10s. 6d. To be continued annually.

A COMPARATIVE VIEW of the relative Produce per Acre, and value of Land, in England, Scotland, and Ireland, applicable to various processes of husbandry, and adapted to the different measures prevailing in the three kingdoms. Printed on a card. Price 6d.

FLEMISH HUSBANDRY.

A REPORT on the **AGRICULTURE** of **EASTERN** and WESTERN FLANDERS, made under the authority of the Farming Society of Ireland. By the Rev. THOMAS RADCLIFF, Secretary to the Society, and Author of the Surveys of Wicklow and Kerry, &c. &c. with Twelve Plates, 8vo. 13s.

In this Work, the Reporter Surveysthe Provinces of Flanders in Eleven Divisons, and gives minute details of the peculiar Husbandry practised in each The work, among other matter generally interesting to the British Agriculturist, comprises the chief varieties of Flemish Rotations according to the variations of the Soil, and in treating of the quantity of Seed and Produce, as well as the application of Manure, the Weights and Measures are conformably reduced to the standards of Great Britain and Ireland. The Appendix contains a mass of information respecting the culture and preparation of Flax, Madder, Woad, and Indigo; the production of Sugar from Mangel Wurzel; Management of the Dairies in Holland: Analysis of Soils, &c. &c The whole illustrated by Twelve Plates, descriptive of Machinery, Farm-buildings, Implements, &c. &c.

HUSBANDRY IN FRANCE.

A SKETCH of the AGRICULTURE of a DISTRICT in the SOUTH of FRANCE. Translated from the French of the BARON PICOT DE LA PEYROUSE. To which are added, Notes by an English Agriculturist, 8vo. 6s.

This Work obtained the Prize offered by the Royal Society for Promotion of Agriculture at Paris, and is interesting from the details which it exhibits of the System of Husbandry pursued by our Neighbours. The Notes that accompany the Work, descriptive of peculiar Manners and Customs, and relating to many topics of great political and commercial interest, are supplied by an intelligent observer.

HUSBANDRY IN CANADA.

FACTS and OBSERVATIONS respecting CANADA and the UNITED STATES of AMERICA; affording a comparative view of the Inducements to Emigration presented in those Countries. To which is added, an Appendix of Practical Instructions to Emigrant Settlers in the British Colonies. By C. F. GRECE, Member of the Agricultural Societies of Montreal and Quebec, author of Essays on Canadian Husbandry, &c. 8vo. 7s.

IMPROVEMENT OF STOCK.

Sir JOHN SEBRIGHT, on the MODE of IMPROVING the BREEDS of DOMESTIC ANIMALS, in a Letter to Sir Joseph Banks, K. B. 8vo. 2s. 6d.

MORLAND'S GENEALOGY of the ENGLISH RACE HORSE; with remarks on the present system of Breeding Blood Horses for the Turf, &c. and instructions for the management of Brood Mares and Colts. With a plate, 8vo. 12s.

LAYING OUT FARMS AND FARM BUILDINGS, PLEASURE GROUNDS, GARDENS, &c.

OBSERVATIONS on LAYING OUT FARMS in the Scotch style, adapted to England, and remarks on the importance of the Berwickshire system of management to the general improvement of landed property. By J. C. LOUDON, F. L. S.—This work is illustrated by Thirty-eight Plates, descriptive of Farm Buildings, machinery, plans for Road-making, embanking, and other rural improvements, in villas, gardens, farms, and pleasure grounds. Folio, 3l. 3s. or with coloured plates, 5l. 5s.

A COLLECTION of DESIGNS for RURAL RE- TREATS, COTTAGES, and VILLAS, principally in the Gothic style of architecture, with their plans laid down to a scale, and other appendages, including copious descriptions. By JAMES MALTON, Architect. 4to. price 2l. 12s. 6d.

This Work contains thirty-four plates and plans of great beauty and variety, adapted to different purposes, and the descriptions, which are full and complete, form a useful body of information on the design and execution of Rural Buildings, according to the object for which they are intended.

DESCRIPTION of an IMPROVED METHOD of DELINEATING ESTATES, combining mathematical accuracy with the beauties of landscape painting, and representing not only the Estate itself, but the adjacent country in panoramic perspective. By T. HORNOR, Land Surveyor, 8vo. with plates, 10s. 6d.

DEARN'S HINTS on an IMPROVED METHOD of BUILDING, applicable to general purposes.

The object proposed is to render External Walls, whether of Cottages or more important Buildings, dry and durable, consistent with economy. To which are added, observations on the USE of SAND STONE and BRICK. 8vo. with plates, 4s. 6d.

LOUDON'S ESSAY on the CONSTRUCTION and ADVANTAGES of PAPER ROOFS, recommended for elegance, durability, and economy, in agricultural and other suitable buildings; with a plate. Second Edition, 2s.

HINTS on the FORMATION of GARDENS and PLEASURE GROUNDS, with designs in various styles of Rural Embellishment, comprising plans for laying out Flower, Fruit, and Kitchen Gardens, and the arrangement of Glass-houses, Hot-walls, and Stoves, &c. adapted to Villa Grounds, from one perch to 100 acres, 4to. with 20 plates, price 2l. 8s.

DIRECTIONS for CULTIVATING the CRAMBE MARITIMA, or SEA-KALE, for the use of the table. By WILLIAM CURTIS, with a coloured plate, 8vo. price 2s. 6d.

THE FLORIST'S DIRECTORY, or TREATISE on the CULTURE of FLOWERS, BULBOUS ROOTS, &c. including a Dissertation on Soils and Manures, by JAMES MADDOCK, Florist; enlarged by CURTIS. New Edition preparing.

EMMERTON'S TREATISE on the CULTURE and MANAGEMENT of AURICULA, POLYANTHUS, CARNATION, &c. with particular directions for preparing the most improved composts, raising new varieties from seed, &c. the whole founded upon thirty years experience; with Figures of Auriculas, &c. Second Edition, enlarged, price 10s.

An ESSAY on SOILS and COMPOSTS, and the Propagation and Culture of Ornamental Trees, Shrubs, Plants, and Flowers. By THOMAS HAYNES, Nurseryman, describing the nature and properties of each, with directions for procuring the ingredients of the most perfect composts, and instructions to make substitutes where the ingredients cannot be obtained. 12mo. 5s.

PONTEY'S PROFITABLE PLANTER; a Theore-tical and Practical Treatise on Planting in every Description of Soil and situation, 8vo. with a plate of implements for planting, invented by the Author. Fourth Edition, 10s. 6d.

TULL'S HUSBANDRY.

TULL (JETHRO) HORSE-HOEING HUSBAN-DRY; or an Essay on the Principles of Vegetation and Tillage, designed to introduce a New Method of Culture, whereby the produce of the Land will be increased, and the usual expenses lessened, 8vo. *A New Edition preparing for the Press.*

HARDING'S NEW CATALOGUE of Books on Agriculture, Planting, Gardening, and Rural Affairs in General. Price 1s.

C. Green, Printer, 15, Leicester Street, Leicester Square.

A

TREATISE

ON

SPINNING MACHINERY;

ILLUSTRATED WITH

PLANS OF DIFFERENT MACHINES MADE USE OF IN THAT ART,

FROM

THE SPINDLE AND DISTAFF OF THE ANCIENTS TO THE MACHINES WHICH HAVE BEEN INVENTED OR IMPROVED BY THE MODERNS.

WITH SOME PRELIMINARY OBSERVATIONS, TENDING TO SHEW THAT THE ARTS OF SPINNING, WEAVING, AND SEWING, WERE INVENTED BY THE INGENUITY OF FEMALES.

AND A

POSTSCRIPT,

INCLUDING AN INTERESTING ACCOUNT OF THE MODE OF SPINNING YARN IN IRELAND.

By ANDREW GRAY,

AUTHOR OF THE PLOUGHWRIGHT'S ASSISTANT, AND EXPERIENCED MILL-WRIGHT.

EDINBURGH:

PRINTED FOR ARCHIBALD CONSTABLE AND CO. EDINBURGH; AND LONGMAN AND CO. LONDON.

1819.

John Moir, Printer.

TO

THE RIGHT HONOURABLE

LADY DUFFERIN.

Madam,

In the following humble attempt to trace the progress of the art of forming a thread, from the simple mode of the spindle and distaff of the ancients, to the more perfect machines which have been invented by the moderns, I have found that the invention of spinning, and the elegant improvement of forming the produce into various fabrics, for comfort and ornament, to which the civilization and improvement of mankind in all countries are so greatly indebted,—have been uniformly ascribed to the ingenuity of the fair sex.

In soliciting your Ladyship's patronage and protection to this little Work, I am guided solely by the benevolent and patriotic attention you have condescended to bestow on the improvement of this one of the earliest of the Arts, and which, allow me to say, places your Ladyship, in modern times, in the same exalted rank as those Ladies deservedly distinguished by the ancients.

Permit me only to add, that your Ladyship will find, in the review of the modern machinery, that it is only applicable to the coarser fabrics in the linen manufacture, and not likely ever to supersede the fine spinning by the hand of females, who, in Ireland, as in other parts of the British Empire, have been always celebrated for their excellence and industry in this truly domestic manufacture.

With the greatest respect,

I have the honour to be,

MADAM,

Your Ladyship's

Devoted humble Servant,

ANDREW GRAY.

Edinburgh, January 1819.

INTRODUCTION.

The object of the following Treatise is to exhibit the construction of several implements which have been long employed by the people of Great Britain, in preparing materials and spinning them into yarn,—from the flat stone and wooden mallet, the spindle and the distaff, which the ancients made use of for these purposes, to various machines which have either been improved or invented by the moderns.

The art of spinning is, without dispute, of very remote antiquity, though we can say nothing certain with respect to its origin, by whom this necessary art was invented, or in what country it was first discovered. Indeed, there are several nations who claim the merit of this valuable invention. The Egyptians say it was Isis who invented, and also taught them the art of spinning.* The Chinese give the honour of this discovery to the consort of their emperor Yao.† It may, however, be observed on this subject, that the traditions of almost all nations ascribe the honour of inventing the arts of spinning, weaving, and sewing, to the ingenuity of the fair sex. The Lydians ascribed this discovery to Arachne,‡ the Greeks to Minerva, § the Peruvians to Mamacella,¶ wife to Manco-capac, their first sovereign. It was also to ladies that the Greek and Roman writers attributed the invention of the needle,|| and the art of spinning the silk of certain insects, and weaving it into stuffs.** We cannot, however, determine whether these tra-

* Martcapella, l. 2. p. 32. † Martini Hist. De la Chine, t. 1. p. 61.

‡ Ovid. Metam. l. 6. *imt.* Plin. l. 7. sect. 57. p. 414.

§ See Part 2. B. 2. sect. 2, c. 2. ¶ Hist. Desineas, t. 1. p. 22. and 51

|| Higin. Fab. 274.

** Arist. Hist. Anim. l. 5 19. p. 849. Plin. l. 11. sect. 26. p. 604. Isidor. Orig. l. 14. c. 6.

ditions be founded in fact, or have only arisen from these particular arts having, in all ages and countries, been allotted to the fair sex.

Be this as it may, doubtless the art of spinning was known, and also practised in very early times. Whoever was the inventor, without doubt it is a most valuable discovery; for by this simple art any fibrous material, as flax, wool, cotton, silk, hemp, &c. can easily be united into one continued thread of any length, and at various degrees of thickness. Hence, from the smallest thread for sewing the finest apparel, to a cable or anchor-rope which holds fast the largest ship of war, spinning has been from time immemorial applied with success. This truth is mentioned by several profane historians at a very remote period; but though it be evident that the art of spinning was practised in very early times, yet we can say nothing certain concerning the manner in which mankind first made use of yarn in fabricating stuffs for raiment. It is mentioned by Homer, that

Penelope, in order to evade the decision of her choice of a husband, told her suitors, that she would make her choice as soon as she had finished a web to wrap round the body of Laertes, the father of Ulysses; and, in order to gain time, she always unravelled in the night the part she had weaved through the day. We should be inclined to imagine this web was some kind of net-work, and that perhaps weaving by warp and woof had not then been invented. Probably, before mankind discovered this valuable art, they would adopt various methods, make many essays, and prepare a variety of fabrics,* as tresses, net-work, &c. till by degrees they found out the web by warp and woof, which, unquestionably, is one of the most useful inventions of which mankind are in possession; for it is by means of this art that various sorts of yarn can be wrought up into warm, commodious, and beautiful garments. Indeed, there are few inventions which have displayed so much ingenuity, or

* See Lucret. l. 6. v. 1349, &c.; Braun. de Vestitu Sacerdot. Hebr. n. 253. p. 250, &c.

done more honour to human genius than that of weaving yarn in o uch an amazing number of different kinds of stuffs as are to be found in all civilized nations Perhaps civilization could not well have proceeded without spinning and weaving, at least we may suppose that navigation and commerce could not have gone forward without these arts. For, besides clothing and ornamenting the human body,—if we consider the commerce of Great Britain and all civilized nations, what could have been substituted for those immense canvas sails by which the largest ships are wafted over the seas to every shore, however distant?

A great many conjectures, however, might be formed about the origin of weaving yarn into cloth.* We might say with an ancient author, that mankind probably owe the discovery of this art to the spider.† They, perhaps, would take notice of the manner in which this insect warped its web; they might

* By weaving, is meant all sorts of works that are made in the Loom.

† Democritus, apud Plut. t. 2. p. 974.

observe how it conducted the threads by the weight of its own body, and thus composed the web of warp and woof. But, without mentioning the various hypotheses which might be formed on this subject, we may suppose that the idea of the web by warp and woof first struck mankind from the inspection of the inner bark of certain trees. Some of these, it is said, bear a very great resemblance to a web; the fibres, it appears, are interwoven, and cross each other almost at right angles. * The manner, therefore, in which the filaments of these barks are disposed, might possibly give the first hint of the web by warp and woof.

Be this as it may, it is evident, that before people could either spin the yarn or weave it into cloth, they must first have invented instruments for preparing the materials, and also machines with which to spin these materials into threads of yarn. The spindle and the distaff perhaps were the chief instruments generally made use of by the ancients in spin-

* It is reported, that trees of this kind grow in the Indies.

ning; and these simple tools are made use of by some people in North Britain at this day. However this may be with respect to spinning, we are still more ignorant as to the form of the weaving instruments which the ancients employed in fabricating stuffs for raiment. If we reflect on the vast number, and great variety of machines, which are now made use of by the inhabitants of Britain in weaving their fine stuffs, we can scarcely believe that the people in these first ages could form any thing like them. Yet it is easy to imagine they might dispense with them, if, instead of viewing our own complicated machines, we observe the simple ones in several remote nations.

It is reported, that the inhabitants of the greater India, Africa, and America, at present weave stuffs with a few very simple instruments. The people in these ancient ages might probably do the same. Though the inhabitants of the above-mentioned countries are certainly strangers to many branches of knowledge which the artizans in Great Britain and Ireland possess, yet the fineness and beauty of their stuffs cannot be sufficiently admired. We can say nothing certain with respect to the construction of the instruments or machines which the people in these countries employ in spinning their fine yarn, or if they use any other than the simple spindle and distaff. We are informed, however, that in weaving they use no other instruments but a shuttle and a few small pieces of wood.

In this age, however, and particularly in Great Britain and Ireland, the arts in general are brought to a considerable degree of perfection, especially that of spinning fine yarn, and weaving it into a variety of beautiful stuffs for clothing, which contribute not only to the convenience and comfort, but also to the ornament of the human body.

Indeed the account we have of the manner in which mankind were clothed before spinning and weaving by warp and woof was invented, gives us some idea of that ignorance which prevailed in those early times with respect to clothing, and is also an indisputable proof of the great advantages which the moderns enjoy in this respect above the ancients.* Their garments discovered neither art nor industry; they used such as nature presented, and needed the least preparation. Some people covered themselves with the bark of trees, others with leaves, herbs, or bulrushes, rudely interwoven.† The ignorance of some savage nations at this day presents us with a model of these ancient customs.‡ The skins of animals, however, seem to have been the most generally made use of for garments in these primitive ages.

Indeed, the skins of animals are not naturally adapted to form an exact and commodious covering for the human body; mankind, therefore, would probably begin very early to search for proper materials of which to fabricate convenient and warm clothing to cover their bodies, without re-

* Letter, Edif. t. 9. 420. Hist. Gen. des Voyag. I. 3. p. 184. Voyag. de Damp. t. 4. p. 232. 233. Hist. des Incas, I. 2. p. 77. Voyag. des de Lerg. p. 275. Nouv. Relat. de la France Equinox. p. 125.

† Strabo. l. II. p. 781. Senec. c. 90. p. 40. C. Hist. des Incas. t. v. Letter. Edif. t. 2. p. 189. Ext. des Histor. Chinois, p. 3.

‡ Voyage de Dampier, I. p. 141. Voyage des Rolland, t. 4. p. 506 and 231.

straining the activity of their motions. The wool and hair of some animals were perhaps the first materials most generally used for making garments. There are, however, several plants, such as flax, hemp, cotton, &c. which are very proper for this purpose. Doubtless, mankind would soon begin to make raiment of cotton, as it requires very little preparation, and has a great resemblance to fine wool. But it is evident that flax, hemp, and other filamentous plants are not so easily prepared as wool and cotton, because the fibrous parts of these plants must first be disengaged from the bark and all other refuse, before they can be spun into yarn, or wove into cloth. In order to this, it is necessary to water, break, switch, or scutch, and heckle them. Yet it cannot be doubted that robes made of linen were used in very remote times. The working up of these materials, however, has been the object of incessant study and reflection. To these earnest inquiries and successive improvements we are indebted for that prodigious number of various kinds of stuffs which are in use among all civilized nations. Linen, in particular, has been used for garments both in Britain and Ireland from time immemorial, and is still made use of in these countries, not only for apparel, but also for a great variety of other purposes. It is obvious, however, that the materials whereof linen is fabricated, must be properly prepared before they can be spun into yarn. Perhaps, therefore, it may be proper, before we begin to describe spinning machines, first, to give some account of several instruments that have been long employed for that purpose, and also of those superior machines which have been more recently invented or improved.

PLATE 1.

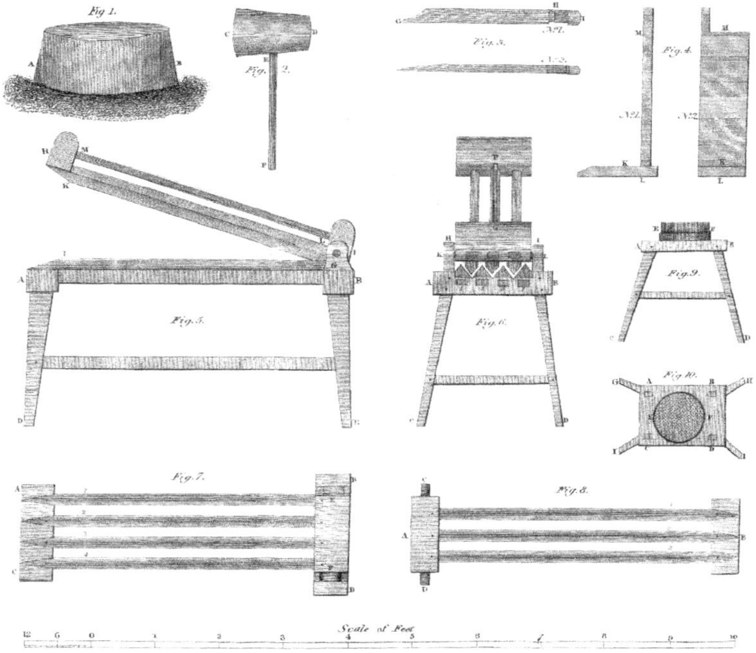

T. Clerk Sculp.

EXPLANATION

OF

THE PLATES.

PLATE I.

Fig. 1. represents a large stone, with a smooth upper surface; a small parcel of the rough plants is laid upon the surface of this stone, and beat till the hard parts are broken and the fibrous parts softened.

Fig. 2. CD, represents a wooden mallet, in which is fixed the handle EF. With this mallet a person beats the fibrous plants, and breaks the hard refuse; the plants, thus softened by repeated strokes of the mallet, are then divided into small parts, and switched, or scutched, with a thin broad piece of wood, called a *swingle-wand*.

Fig. 3. No. 1. GHI, is a profile of the *swingle-wand*, and No. 2. is an edge view of the same instrument.

Fig. 4. No. 1. ML, is a cross section of a board termed the *swingle-stock*, the lower end of which is fixed into the block of wood KL. The manner of working with this instrument is by a person seated upon a stool, or chair, with one foot placed upon the block at K, to hold the instrument firm; then taking a parcel of the broken or beat plants into one hand, holds it over the top of the *swingle-stock* at M (see M, Fig. 2.) then grasping in the other hand the *swingle-wand* GI, at H, Fig. 3. No. 1. and 2. This instrument is thick on one edge, thin on the other, and a little rounded, to prevent it from cutting the materials, and with it the parcel placed over the swingle-stock is struck, till the loose rough refuse is detached from the small fibrous, tough, or soft parts. Thus, by these different operations, the fibrous plants are prepared for the heckle, by which they are made ready to be spun into yarn.—The next machine used in Britain for breaking rough flax, or any other fibrous plants, seems to be what was termed a *brake*. This, no doubt, was superior to the stone and mallet, because by it the plants are broken with more expedition, and also more perfectly.

Fig. 5. Profile of a machine for breaking flax, or any fibrous plants.

ABDE, represents a strong wooden frame, into which are fixed four longitudinal spars, as IG, of hard wood, made thin to their upper edge, and a little rounded, to prevent them from cutting the plants.

HI is a moveable frame, containing three spars, (see AB, Fig. 8.) This upper frame has liberty to turn upon two pivots, (see CD, Fig. 8.) in the frame at G; and when using, the three upper spars fall into the openings between the lower ones, (see 2, 3, 4, Fig. 7.)

ML is a rod of wood, fixed into the blocks H and I, by which a person moves the frame H upwards and downwards with one hand, at the same time holding a parcel of the rough plants across F G, the lower spars; so that by the repeated strokes of the upper frame, HI, the rough plants are broken and prepared to be scutched, or the loose refuse detached from the small fibrous parts, by means of the swingle-stock and swingle-wand.

Fig. 6. is a cross section of the same machine.

ABCD represent the frame, wherein are fixed the spars 4, 5, 6, 7.

FG is the upper frame, containing the spars 1, 2, 3, fixed in the block HI, in which are two pivots having liberty to turn in the frame at K and L. There is a wooden rod fixed in the blocks E and G, by which a person moves the upper frame FG upwards and downwards, when breaking the hard parts of any fibrous plants.

Fig. 7. Plan, or bird's-eye view, of the same machine.

ABCD repsesent the lower frame, in which are fixed the longitudinal spars 1, 2, 3, 4.

EF are two uprights, fixed in the lower frame at B and D, into which the pivots of the upper frame turn when this machine is employed. (See G, Fig. 5. and CD, Fig. 8.)

Fig. 8. Plan of the moveable, or upper part of the brake.

AB represent two blocks of wood, into which the longitudinal spars 1, 2, 3, are fixed. CD are the two pivots upon which the upper frame turns, when its fore-end B is moved upward or downward. (See H. Fig. 5.) The hard part of the flax is thus sufficiently broken and softened by repeated strokes of the upper frame, which is more expeditious than the stone and the mallet ; the work is also performed in a more perfect degree. Still, however, the loose refuse must be switched off, either with the stock and swingle-wand, or by some other instrument, in order to prepare the broken materials for the heckle, which is the last operation in preparing flax, hemp, &c. to be spun into yarn. As the before-mentioned instruments for breaking and scutching flax, hemp, or any other fibrous plants, are wrought by the hand, of course it is very laborious and also tedious work ; but perhaps there were no better machines for that purpose in those early days. Indeed, the time is not very remote since there was not any proper machines in Scotland for breaking and switching flax, hemp, &c.

Fig. 9. A section of the ancient heckle, wherein AB and CD represent a table upon which the heckle was fastened. EF is the heckle as placed upon its table.

Fig. 10. Plan, or bird's-eye view of the table, with the heckle EF fixed upon it. This heckle was a round block of wood, into which were fixed small iron spikes, which were placed in different circles, at a small distance from each other. The heckle is the last machine employed for removing all refuse from flax, hemp, and other fibrous plants; and this operation of heckling, till of late years, was generally performed by females. GHIK are the feet whereon this table stands when using.

Thus have we given figures, with a short description of different instruments which were long employed in preparing fibrous plants to be spun into yarn, and shall next proceed to give some account of that valuable machine, the flax or lint-mill, driven either by water, wind, cattle, or steam.

PLATE II.

The next machine invented for breaking and scutching flax, hemp, or other fibrous plants, seems to be what is termed the lint or flax-mill. This, doubtless, is a very complete machine for that purpose, because the operation is performed in a superior manner, with ease and expedition. Therefore we have given a plan, elevation, and profile of this very useful machine, which may possibly be acceptable to some readers, as it is without dispute the most complete machine for breaking the hard parts, and separating the rough refuse from flax, hemp, &c. which has hitherto been introduced into North Britain, or perhaps into any other country. It is therefore somewhat surprising that a machine for these important purposes, turned by water, was not introduced into North Britain at a more remote period; for it is doubtful if there was such a machine in that country a century ago, and probably there were very few before the year 1730, about which period two flax or lint-mills driven by water were erected in the county of Perth; one of them at Rait, in the parish of Kilspindie, and the other

PLATE II.

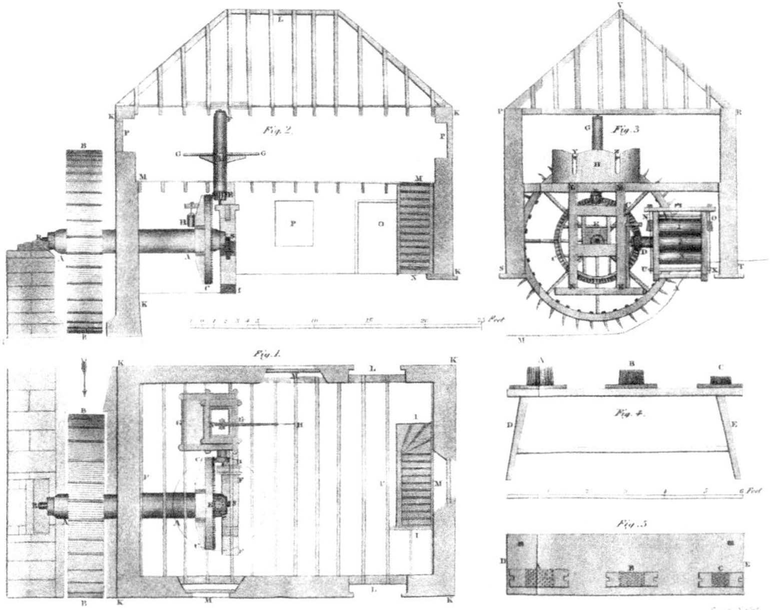

at Killamacker, in the parish of Comrie. These machines, however, were found so very useful in breaking the hard parts, and scutching or switching off the loose refuse from flax, hemp, or other fibrous plants intended to be manufactured into yarn, that in the year 1802 there were upwards of 120 in the county of Perth, and more than 400 in North Britain; so rapidly the number of these machines was increased between the above periods.

FIG. 1. Plan, or bird's-eye view of a machine for breaking and scutching flax, hemp, &c.

AA represent the axle, or shaft, upon which is fixed the great wheel BB, termed the water-wheel, having placed on its circumference float-boards, or buckets, to receive the water which turns it round to drive the breaking rollers and scutching machine. CC is a wheel fixed upon the axle AA, containing 102 teeth, which turns the pinion D, having 25 teeth, which is fastened upon one end of the middle breaking-roller. E represents a pinion containing 10 teeth, which is also turned by the wheel CC, and is placed upon the lower end of an upright shaft. (See E, in Fig. 2.) FF, the frame in which the pivot S of the axle A turns when in motion. GG, the frame in which the breaking-rollers revolve when employed in crushing any hard fibrous plants.

HN is a lever, having liberty to turn upon an

iron bolt at G; its shorter arm N is attached to a frame, in which the pivots of the upper roller turn; so that by pushing down the longer arm H this roller is raised clear of the middle one at any time when necessary. The circle CD and CF is that part of the loft where the box, or frame in which the scutchers revolve, is placed. KK and KK are the walls of a house to contain the breaking rollers and scutching machine.

PP is the joists, or frame of the loft upon which the scutching-box is placed. II, the stair leading up to the loft. LL are doors in the side-walls. (See O, Fig. 2.) MMM, windows to light the house and machinery.

Fig. 2. Elevation of the house and machinery.

AA, represent the axle or shaft, and RS the pivots upon which it turns; BB, the great wheel fixed upon this axle, on the rim of which is placed float-boards to receive the water by which it is turned round. Upon the axle AA is also fastened the wheel CC, containing 102 teeth, or cogs to turn the pinion E, having 10 teeth, which is fixed upon the lower end of the axle FE. There are also two arms that pass through mortises in this axle, at right angles to each other, and upon these arms are fastened, with screwed iron bolts, the scutchers GG, of which there are four, to clear the rough refuse off the broken flax or fibrous plants, which are opposed to them as they revolve. EI is a strong wooden frame, in which the pivot S of the axle A, and also the pivot in the lower end of the axle EE, turn when in motion. H is a weight suspended by a rope, the other end of which is attached to a cross-bearer below the sliders, in which the pivots of the lower roller turn, so that by these weights this roller is kept close up to the middle one. (See NOUX, Fig. 3.) KK and KK are the walls of the mill-house; MM is the loft or second floor; MN a stair leading up to the loft; KKL the couples, or framing of the roof; O the door; (See LL, Fig. 1.) PPP are windows in the walls to light the house and machinery.

FIG. 3. Profile of the same Machine.

AA represents the great wheel fixed upon the shaft or axle E, and upon the circumference of this wheel are fixed buckets, or float-boards, to receive the water by which it is driven round. CC is a wheel fastened upon the same axle A, containing 102 teeth, or cogs, to turn the small wheel D, in which are 25 teeth, which is fixed upon one end of the middle breaking roller, No. 2. This roller is fluted, or has longitudinal teeth all round its circumference. These teeth are of an angular form, being broad at their base, and sloped equally on both sides towards their points, which are a little rounded; by this means they are prevented from cutting the rough flax, or other fibrous plants, in passing through between the rollers. The other two rollers, Nos. 1. and 3. are also fluted or have longitudinal teeth of the same form and size with those in No. 2. or the middle roller; and by its teeth taking into the flutes of the rollers No. 1. and 3,. they are both driven round when the great wheel AA is in motion. The rough plants being made up into small parcels, are introduced by the hand betwixt the rollers No. 1. and 2. and are drawn round the middle roller, the back part of which is inclosed by a curved board, that turns the rough plants out through between the middle and under roller; they are then put in again betwixt the rollers No. 1. and 2. going round the same course until their hard parts be sufficiently broken or softened, and thus prepared for the scutching machine, by which the loose refuse is cleared off from the useful part of the plant. The pivot of the roller No. 2. turns in a bearer fixed in the frame; and the pivots of Nos. 1. and 3. turn in brass bushes fixed in wooden sliders, which have liberty to move up and down into grooves in the frame NO and UX. The under roller, No. 3. is kept close up to the middle one, by the weights NO, suspended on two ropes going over sheeves or whorls in the roller frame NOUX, their other ends being fastened to a cross bearer below the sliders in which the pivots of No. 3. turn. The weights NO must be considerably heavier than the roller No. 3. with its sliders, in order to press its teeth in between the flutes of No. 2. by which it is turned round, to break the rough plants, as they pass through between these rollers. The whole weight of the upper roller, or No. 1. presses upon the rough plants, as they pass through betwixt it and No. 2., or the middle roller; but if more weight is required than the roller with its sliders, there is a box made of boards placed upon the sliders in which the pivots of No. 1. turn, and into this box a parcel of stones, or any heavy metal, may be put, and by this means more or less weight can be added to the upper roller when found necessary. FG the perpendicular axle or shaft, upon the lower end of which is fixed the pinion F, having 10 teeth, which is turned by the wheel CC. In the axle G are also fixed two arms which cross each other at right angles, and upon these arms the scutchers are fastened with screwed iron bolts. (See GG, Fig. 2.) The axle G, with the scutchers, revolves within the frame H, or what is termed the *scutch-box*. YZ represent two of the openings, of which there are six, and often eight, in the circumference of this box; at these openings the broken materials are introduced by the hand, and struck by the scutchers as they revolve, and thus the rough loose refuse is cleared off from the broken flax or fibrous plants; TM the course, or fall of water, which drives the great wheel A, the wheel CC being fixed upon the same axle to turn the breaking rollers and scutching machine. PRST are the side walls of the mill-house; PRV the couples or framing of its roof.

Fig. 4. DE represent a frame upon which the heckles AB and C are fixed; the teeth of these heckles are placed in straight lines, which seems to be an improvement on the old practice of placing the heckle teeth in circles.

Fig. 5. ABC represent the same heckles fixed upon the board or table DE. (See ABC, Fig. 4.) By these instruments flax or other fibrous plants are cleared of refuse, and thus completely prepared to be spun into yarn.

PLATE III.

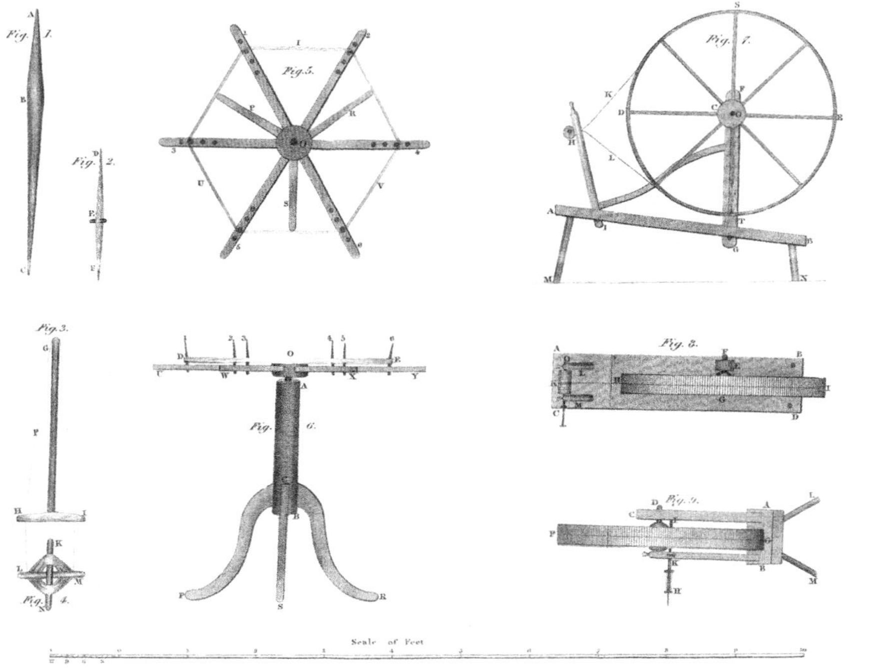

PLATE III.

Plan, Elevation, and Sections of several Instruments and Machines which the Ancients employed in Spinning fibrous materials into Yarn.

Fig. 1. ABC represent the distaff, or a rod of wood upon which the materials to be spun are placed; these materials are rolled round upon that part of the distaff from A to B, and kept fast by a ribband or garter tied round it. The lower part, BC, is kept fast with a girdle or belt tied round the body of the spinner; by this means the distaff is kept steady whilst the person is employed in spinning.

Fig. 2. Is the spindle, which being turned quickly round by the spinner's right hand, twines or twists the thread drawn from the materials on the distaff by the spinner's left hand. The thread thus twined, or twisted, is rolled upon the spindle between D and E. A little below E is placed upon this spindle a small wheel called a *whorl*, generally made of stone, which, acting as a fly, makes the spindle revolve uniformly, as it turns on its lower point F, upon the floor or pavement. By this means flax, or any fibrous materials, are regularly drawn out into a thread of yarn.

These seem to be the most simple instruments that have been employed in spinning, and perhaps they are also the most ancient; and there are some persons in North Britain who make use of these simple instruments in spinning at this day, 1817.

Fig. 3. Profile of an instrument termed a *hand-reel.*

G represents a wooden rod, upon each end of which is fixed a small piece of wood, placed upon the rod at right angles to each other; G is an end view of one cross, and HI the length of the other. A person holding this rod in one hand, as at P, with the other hand conveys the thread of yarn from G over the cross piece at H, and again from H over the cross piece at G, then from G over the cross piece at I, from I to G, and so on, 120 times, which number makes one *cut*, and is tied up by itself. Twelve of these *cuts* make a hank or hesp, and four of these hanks make what is called a spindle.

Fig. 4. Section of the same machine, with the yarn placed upon it.

KLMN represent the pieces of wood fixed at

right angles across each other. The thread when reeling passes from the arm L, over above the arm K, and from that below the arm M, from which it goes over above the arm N, passing again below the arm L, and so on, until the twelve *cuts*, or one hank of yarn, is completed.

Fig. 5. Plan or birds-eye view of a machine upon which the hank of yarn is placed, and from which the threads are rolled upon a pirn or bobbin, (See H, Fig. 9.)

O represents a round piece of wood, in the middle of which is a small hole to take in an iron spike, fixed in the upper end of a wooden pillar, (See AB, Fig. 6.) upon which this machine revolves when put in motion. Into this round block of wood is fixed one end of the arms, Nos. 1, 2, 3, 4, 5, and 6; near the extremity of these arms are several holes to receive small pins of wood, (See 1, 2, 3, 4, 5, and 6. Fig. 6.); round the outer side of these pins is placed the hank of yarn, and from thence rolled upon the bobins, (See H, Fig. 7.) The wooden pins may be placed into the holes of these arms, either nearer to the centre or farther from it, according to the length of the hank of yarn. IUV is the yarn placed upon the arms. PRS are the feet which support this machine. (See CPRS, Fig. 6.)

FIG. 6. Elevation of the same Machine.

AB represent the pillar of wood, into which is fastened a small iron spike, and upon it the block O revolves with its arms UWXY; into these arms are placed the wooden pins 1, 2, 3, 4, 5, & 6.

DOE, the hank of yarn placed upon the outside of the pins; by this means the hank is extended, and in a great measure prevented from raveling, when the arms, UWXY, upon which it is placed, revolve round their centre at O, whilst the threads of yarn are rolling upon the pirn. (See H, Fig. 9.) CPRS are feet fixed into the pillar AB, upon which the machine stands when employed.

FIG. 7. Profile of a machine termed the pirn-wheel, by means of which the yarn is rolled upon the pirn or bobbin.

AB represent a plank of wood, and FG a wooden pillar fixed into the plank at T; in this pillar, at C, is fastened a round piece of iron or wood upon which the wheel DES revolves. (See F, Fig. 8. and D, Fig. 9.) CO is a round piece of wood, and in its middle is a hole that takes in the axle O, upon which this wheel revolves when in motion. Into the block C are fixed eight spokes, or arms, and on the extremity of these arms is fastened a thin broad piece of wood that forms the wheel DEST; upon this wheel is placed the band or belt KL, which passes round the whorl

H, fixed on the spindle of the pirn. By (See KC, Fig. 8.) this means, when the wheel DE is driven round, the whorle H and its spindle, with the pirn, is also turned round; and of course the yarn is rolled upon the pirn.

HI is a wooden frame, fixed at I into the plank AB; and in this frame the spindle, with its whorle and pirn revolve, (See OKC, Fig. 8.)

MN are two of the feet whereon the machine stands when worked.

Fig. 8. Plan, or bird's-eye view, of the same machine.

ABCD, represent the wooden plank into which is fixed the piece of wood E, and into it is fastened the axle F, upon which axle the wheel HI revolves. The band or belt IH, pass from this wheel round the whorle K, placed upon the spindle OC; on this spindle is also placed the pirn at C, and upon it the yarn is rolled, when the machine is in motion.

Fig. 9. Cross section of the same machine.

AB, represent a plank of wood, and into it is fixed AC, and into AC is placed the axle D, upon which the wheel FG revolves; HI, an iron spindle turned by a band from the wheel FG, and upon this spindle is placed the pirn H.

BK, The frame in which the iron spindle with its pirn revolves, when the wheel FG is turned round. LM are two of the feet that support this machine.

PLATE IV.

Plan, Profile, and Section of the Common Spinning-Wheel.

The spindle and the distaff, no doubt, had been made use of in spinning long before the spinning-wheel was invented. However, as we can say nothing respecting the origin of the spindle and distaff, neither can we say any thing certain in regard to the origin of the spinning-wheel, by whom it was invented, or into what country this useful machine was first introduced. Probably it has undergone several alterations since first invented, and is now brought to a considerable degree of perfection; nevertheless it may possibly still admit of some small improvement.

Fig. 1. Profile of the common One-handed Spinning-wheel.

AB represent what is termed the wheel-*stock*, to which all its other parts are attached; C and D are two of the feet or pillars which support this machine; EF is the wheel or *rim* round which goes the band or belt GH, and passing from this wheel over the whorle or sheave at IE, by this

PLATE IV.

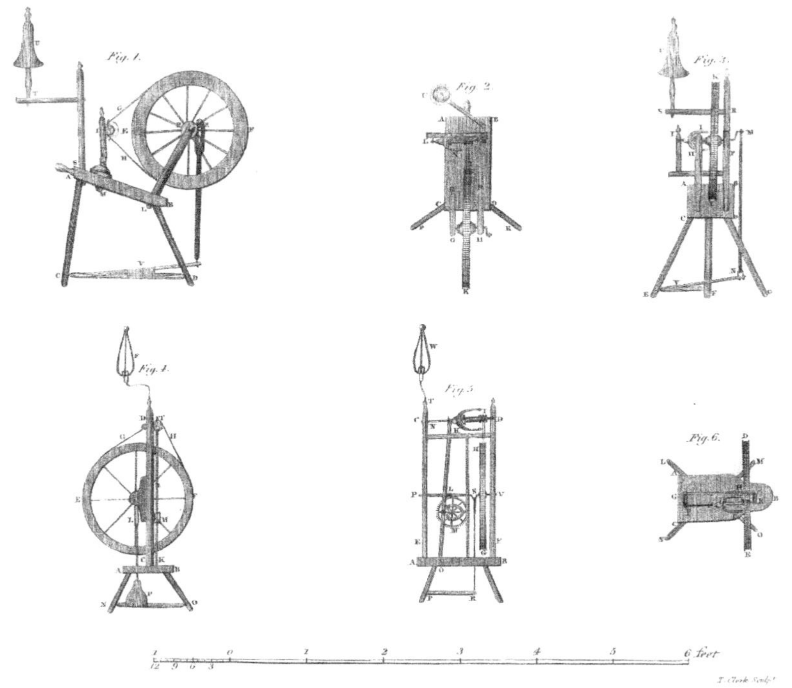

Fig. 1.
Fig. 2.
Fig. 3.
Fig. 4.
Fig. 5.
Fig. 6.
1 0 1 2 3 4 5 6 feet
12 9 6 3
T. Clark Sculp.t

means the turn-spindle, with its heck, or *fly*, and pirn on which the yarn rolls, is turned round.

PL are bearers in which the pivots of the wheel EF revolve when in motion; DR is a small wooden rod, by which motion is conveyed from the frame VCD; and upon this frame, at V, the spinner's foot is placed to turn the wheel, EF, by means of a bend or crank in its axle, and upon this crank is placed the end, R, of the rod RD (see M, Fig. 3.) Thus motion is conveyed to the wheel EF, by a person's foot moving the rod DR upwards and downwards when spinning.

S is a wooden screw, that, when turned over to the left-hand above the band GH, is slackened, to cause the fly and pirn move slower; and if turned over above to the right hand the band is tightened, to make the fly and pirn IE revolve quick, by which means the motion of the spindle, with its heck, or fly and pirn, may be regulated at pleasure. ST represent what is termed the *standard*; TU is a pillar placed into a hole in the *standard* at T; and upon this pillar the materials to be spun are placed.

IM is a frame in which the spindle turns, with its fly, or heck and pirn. (See HI, Fig. 8.)

Fig. 2. Plan, or bird-eye view, of the same machine.

ABCD represent the *stock*, GH are two pillars

fixed into it at EF, and in these pillars at GH the axle of the wheel IK turns. LM is a frame into which the iron spindle revolves with the fly, or heck and pirn, TN. (See HIT, Fig. 3.) I, is the band or belt which pass round on the wheel IK, and over a whorle at M, fixed upon the spindle of the fly and pirn; by this means, when the wheel IK is turned round, the spindle, with its fly and pirn on which the thread of yarn rolls, is also turned round.

S is a wooden screw, by turning of which a little round, the band, I, can be straitened or slackened when found necessary. U is the *rock* or pillar on which the materials to be spun are placed. (See U. Fig. 1.) PR are two of the feet which support this machine when employed.

Fig. 3. Elevation of the same machine.

ABCD represent the wheel-*stock*, HP the bearers into which the axle of the wheel KL revolves. (See PL, Fig. 1.) EFVN, is a frame termed the foot-board, and upon it the spinner places one foot at V, the moving of which conveys motion to the wheel KL, by one end of the rod MN being attached to the foot-board at N, and its other end placed upon the bend or crank M, on the axle of the wheel KL. So that when the frame EFVN is moved by the spinner's foot, then the wheel KL is turned round, and of course the iron spindle, with its fly and pirn, is also turned by a band or belt from the wheel KL. (See GH, Fig. 1.)

IP is a frame in which the iron spindle, with its fly and pirn IHT, revolve. PR, the *standard* placed into the *stock* at B; and in the cross-piece RS is placed the *rock*, or pillar U, upon which the materials to be spun are rolled, and from it the thread of yarn is conveyed by the spinner's fingers to I, the iron spindle, with the fly and pirn HT, on which the thread of yarn rolls when the machine is in motion. EFG are the feet that support this machine.

Fig. 4. Profile of Mr SPENCE's improved Spinning-wheel.

AB represent a piece of wood termed the wheel-*stock*, into which the uprights, CD, are fixed, and in these pillars the axle or spindle of the wheel EF revolve. (See PV. Fig. 5.) GH is the band or belt which passes over this wheel, and from it over a *whorle*, or pulley, placed upon the iron spindle of the fly and pirns DT; by this means, when the wheel EF is moved round, the iron spindle, with its fly and pirn, is also turned round by the band GH. (See GHIK, Fig. 5.) L is a small rod, one end of which is attached to the foot-board P, and its other end upon a crank or bend in the axle of the wheel EF. (See S, Fig. 5.) AMT is a rod of iron which has liberty to move on a pin at K, and its other end T moves the heck forwards and backwards upon the iron spindle, whilst the thread is rolling upon the pirn. P is the foot-board, which has liberty to move on pivots in the feet at N and O; so that when this board is moved by the spinner's foot, the rod L is moved up and down, by which motion is conveyed to the wheel EF.

FIG. 5. Elevation of the same machine.

A B represent a piece of plank, to which all the other parts of this machine are attached.—C D E F is a frame, the lower end of which is fixed into the plank AB, and into this frame the pivots of the great wheel GH revolve. LM is a wheel fixed upon an iron axle, and in the circumference of this wheel are small teeth, which take in the spirals of an endless screw, fixed at L, upon the axle of the wheel GH; of course every revolution of the screw at L moves the wheel M one teeth forwards, and upon the axle of this wheel is fixed a hart, or eccentrical wheel; so that when the wheel M is turned round by the screw at L, then the eccentrical wheel moves the rod OK, which, having liberty to turn upon an iron bolt at O, the end K carries the fly or heck IK from K to N, with the thread of yarn in its point at I, and when the heck IK arrives at N it returns again towards K, placing the thread uniformly upon the pirn, from one end to the other. This prevents what happens very often, and is so hurtful on the common wheel—all chance of the yarn ravelling when it comes to be reeled; besides this advantage, the spinner never needs to stop for shifting the thread from one tooth in the heck or fly to another, which saves much time and trouble to the spinner.

FIG. 6. Plan, or bird's view, of the same machine.

ABC represent the wheel-*stock*; DE, an edge view of the great wheel. (See EF, Fig. 4.) GH is a frame, in which the axle of the wheel DE, the spindle PR, with the fly HI, and its pirn, revolve; and upon the part P of the spindle the heck has liberty to move forwards and backwards the whole length of the pirn, in order to place the thread equally and regularly from one end of the pirn to the other. LMNO are feet fixed into the stock ABC, and upon these feet the machine stands when employed in spinning. As every alteration in a machine cannot be termed an improvement, therefore this alteration on the spinning-wheel being of a very recent date, we cannot determine whether it is an improvement or not until its merits be fully proved by experience.

PLATE V.

Fig. 1. is a cross section of a Two-handed Spinning-wheel.

AB represent a plank of wood termed the wheel-*stock*; CD is a frame, the lower end of which is fixed into the stock, and in its upper end the iron axle, with the wheel EL, revolve. (See BC and NG, Fig. 2.) CF is the fly or *heck*, with the pirn upon which the threads of yarn are rolled. HIK is a frame termed the foot-board, the pivots of which turn in the two feet H and I; so that if the foot-board be moved by the spinner's foot, of course motion is conveyed to the wheel EL by the rod B, one end of which is attached to the foot-board at K, and its other end placed upon the bend or crank D of the axle which carries the wheel EL. Thus, when the foot-board is moved, then the rod B is pushed upwards and downwards, by means of which the wheel EL is turned round, and conveys motion to the iron spindle N with its heck and pirn FC, on which the thread of yarn rolls as it comes from the fingers of the spinner. NO is the frame in which the iron spindle with the heck and pirn

PLATE V.

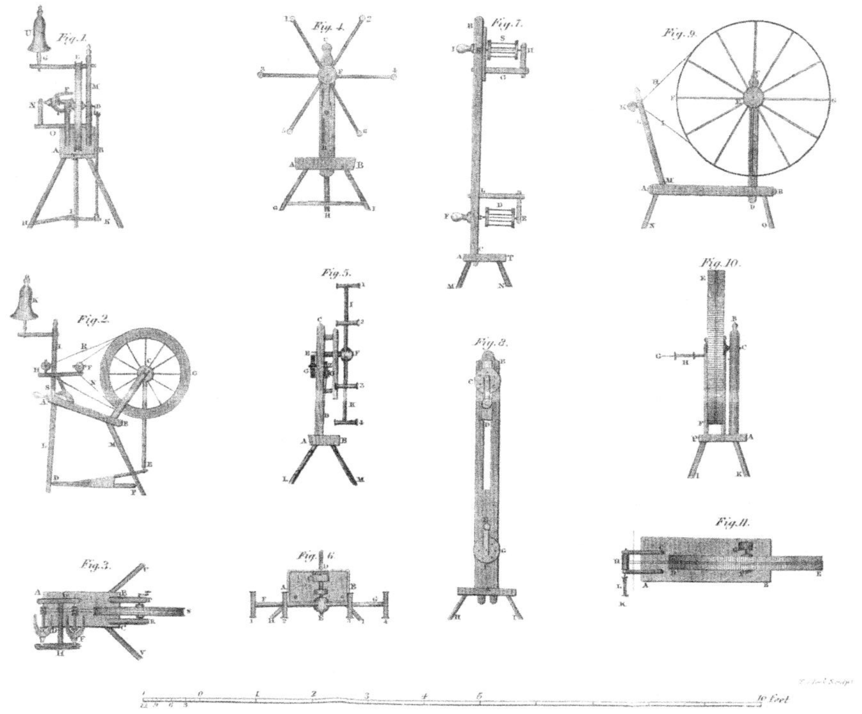
Fig. 1.
Fig. 2.
Fig. 3.
Fig. 4.
Fig. 5.
Fig. 6.
Fig. 7.
Fig. 8.
Fig. 9.
Fig. 10.
Fig. 11.
10 Feet

CF revolve. M, the *standard;* EG a cross piece of wood fixed into it; and in this cross, at G, is placed the *rock* or pillar, GU, upon which the materials to be spun are placed. HIK are the feet on which this machine stands.

Fig. 2. Profile of the same Two-handed Spinning-wheel.

AB represent the *stock;* BC the pillars in which the axle of the wheel FG turn when in motion. CE is a wooden rod fastened at E to the foot-board DEP, and its other end placed upon a crank or bend in the axle of the wheel FG. (See D, Fig. 1.); by this means, when the foot-board is moved, the wheel FG is put in motion. RN are two bands, or belts, which go over the wheel FG; the band N, passing from the wheel G, drives the fly and pirn F, and the band R turns the fly and pirn H, when the wheel FG is put in motion. A is a wooden screw, by turning of which the bands R and N may be made strait or slack at pleasure. IS is the standard, with the *rock* or pillar K, whereon the materials to be spun are rolled. TM are two of the feet that support the machine when employed.

Fig. 3. Plan or bird's eye-view of the same Two-handed Spinning-wheel.

ABCD represent the *stock;* DF is the two flys and pirns placed upon an iron spindle which revolves in the frame GH, when turned by a band from the wheel S, the axle of which turns in the bearers PR. (See BC, Fig. 2.) UV are two of the feet that support the wheel when used in spinning. At what period the Spinning-wheel was introduced into Britain perhaps is not certainly known; but it is recorded that a wheel or machine for spinning yarn was used in London so far back as the year 1681; about which time it appears Thomas Firman, Esq. erected a school for the benevolent purpose of teaching poor children to spin, and it seems the two-handed spinning-wheel was then in use.

Fig. 4. Elevation of the common Reel, by which the number and length of the threads of yarn in each cut are ascertained.

AB represents a piece of wood, into which the upright frame CD is fixed; E is a round wooden block, and into it are fastened one end of the arms or spokes 1, 2, 3, 4, 5, 6; and upon the extremity of each arm is fixed a small piece of wood, (see 1, 2, 3, 4, Fig. 5.) and over the outmost side of these cross heads the threads of yarn are rolled. GHI are three feet, fixed in the block AB, and upon these feet the machine stands.

Fig. 5. Profile of the same machine.

AB is a piece of wood, into which the upright

frame CD is fixed; and in this frame the axle EF, with the arms IK, revolves, with the cross pieces of wood, 1, 2, 3, 4, which represent four of the six arms in the reel, (see 1, 2, 3, 4, 5, 6, Fig. 4.) round the uttermost parts of which the threads of yarn are rolled when the machine is in motion. In this axle, at E, is fixed four round spikes of iron, which take into the teeth of the wheel G, containing 24 teeth; and upon its axle is fixed, at Z, an endless screw, the threads of which take into the teeth of the wheel O, having 20 teeth; so that for every six revolutions of the axle EF, with its arms IK, the wheel G, with the endless screw Z, is turned once round, and of course the wheel O is moved one teeth forwards. In one side of this wheel is fixed an iron spike, by which the spring DO is raised a little up at O; and when the axle EF, with the arms IK, has made 120 revolutions, then the wheel G has made 20 revolutions, at the same time the wheel O has made only one, and immediately the spring at O falls off from the iron spike in that wheel, which makes one cut. Thus 120 turns of the arms IK make 120 threads, each thread measuring 90 inches, or seven feet six inches in length; so that as 120 threads make one cut, so twelve of these cuts make what is termed one hank or hesp, and four of these hesps make one spindle, or 5760 threads round the extremity of the reel, measuring 14,400 English yards, or 43,200 feet. LM are two of the feet or pillars that support this machine when employed.

Fig. 6. Plan or bird's-eye view of the Common Reel.

AB represents a square piece of wood, into which is fixed the frame CT; and in this frame the axle E revolves, and upon it is fixed the centre block ET, into which is fastened the arms FG. 1, 2, 3, 4, represent four of the cross heads fixed upon the extremity of the arms FG, and round on the extremity of these cross heads the yarn is rolled in reeling. DHI are the feet upon which this machine stands.

Fig. 7. Profile of a machine termed the Swifts.

AT is a square piece of wood, into which is fixed the upright BC, and into it is fastened at L the frame EL, and in this frame the lower reel D revolves. F is a wooden handle, screwed into the upright B C; in the inner end of this handle one pivot of the reel D turns, and its other pivot turns at E in the frame L E. The uppermost reel S revolves in the frame K H G; this frame has liberty to move up or down into a long mortice in the upright B C (see D, Fig. 8.), so that the reel S may be easily shifted up or down, and placed at a greater or lesser distance from the reel D, according to the length of the skeen or hank of yarn, which goes over both these reels, and from them the thread of yarn is rolled upon a pirn or bobbin. I is a handle of wood, screwed into the moveable frame at K, so that by turning this handle a little round, the frame K H G can be easily shifted up or down, and secured upon the upright B C at any place that may be found necessary. M N are two of the feet that support this machine.

FIG. 8. Section of the same Swifts.

B G represent a frame of wood fixed at A into a square piece of plank. E is the lower frame, in which the reel F G revolves. C is the uppermost reel, and D the frame in which its pivots turn when in motion, the frame D having liberty to move up or down in its mortice, according to the distance required between these two reels. H I are two of the feet on which this machine stands when at work.

FIG. 9. Profile of a machine termed a Pirn-Wheel, by which the threads of yarn are rolled either upon a pirn or bobbin.

A B is a strong plank of wood, into which is fixed the upright C D; and into this upright is fastened the axle upon which the centre block E, with its wheel F G, revolves. (See C, Fig. 10.) H I is the band or belt that goes upon this wheel, and from it over the whorl K, which is fixed upon an iron spindle whereon the pirn or bobbin is placed. (See G and H, Fig. 10.) L M is the frame in which the iron spindle with its pirn revolves. (See C K and H L, Fig. 11.) N and O are two of the feet that support this machine.

FIG. 10. Cross section of the Pirn-Wheel;

In which A P represent a wooden plank; AB a piece of wood fixed upright into this plank at A, and into the upright, AB, is fastened the axle C, upon which the wheel E F revolves when in motion. G is an iron spindle, whereon the pirn H is placed, and upon this pirn the yarn is rolled. I K are two of the feet upon which this machine stands when using.

FIG. 11. Plan of the same Pirn-Wheel.

A B represent a plank of wood, into which is fixed the frame C D; and in this frame the iron spindle K revolves, having the whorl H fixed upon it, and over this whorl goes a band or belt that passes from the wheel D E; by this means the iron spindle K is turned round with its pirn L, and upon this pirn the yarn is rolled when the wheel D E is driven round. F G is the axle upon which the wheel D E revolves when in motion.

Thus we have given a few Plans, with a short description of several instruments and machines

PLATE VI.

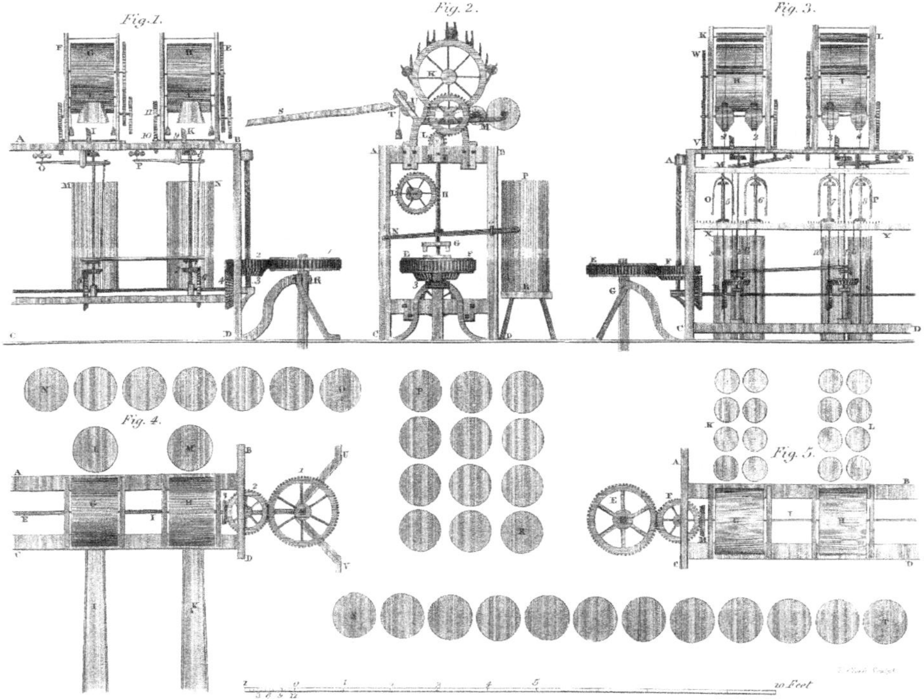

that have been long employed in spinning yarn by the hand; and shall proceed to give drawings of some spinning-machines which have been recently invented, and may be driven either by the hand, by water, by wind, by cattle, or by the power of steam, as may be found most convenient, according to the number of pirns, or the power required to drive the machine.

PLATE VI.

FIG. 1. Profile of a Machine for Spinning Flax, &c. into yarn.

ABCD represent a frame of cast metal, in which the preparing machine is placed. EF are two of the frames in which the cylinders GH with their wheels revolve. IK are two spouts, and along each of these spouts is laid a small parcel of the materials to be spun, and introduced in between the rollers LL, by which it is drawn forward, and formed into a narrow flat thread, without any twine or twist. No. 1. is a wheel fixed upon an iron axle R, upon the lower part of which is fixed a drum-wheel to receive a leather-belt coming from a drum-wheel of any engine in motion to drive this machine. No. 2. is a wheel fixed upon an iron spindle, and is driven round by the wheel No. 1. Upon the spindle of No. 2. is also fixed the conical wheel No. 3. to turn the wheel No. 4. upon the axle of which is placed the two conical wheels Nos. 5. and 6. to turn two pinions that are fixed on the lower parts of two iron spindles, as No. 7. which is turned by No. 6; and upon the iron spindle of the wheel No. 7. is fastened the pinion No. 8. to turn the conical wheel No. 9. upon the axle of which is fixed the wheel No. 10. to turn the wheel No. 11. and convey motion to the cylinders GHLL when the machine is in motion. O and P are two levers, by which the cylinders are stopt at pleasure. MN are two white-iron tubs, into which the materials are dropt from the cylinders.

FIG. 2. Cross section of the same machine.

ABCD represent the frame in which the wheels and cylinders revolve. EF is a wheel fixed upon an iron axle (see R, No. 1. 2. Fig. 1.) G is a cross, fixed upon an iron spindle, having two branches which drop into holes in the wheel below it; by this means the spindle is turned when the wheel EF is in motion, and can be stopped at any time by raising the lever G, which turns upon an iron bolt fixed in the frame at N. Upon this spindle, at H, is a perpetual or endless screw, that takes into the teeth of the wheel I to turn it round; and upon the axle of the wheel I is fastened a hart that moves the frame with the pirns upwards and downwards; by which means the threads of yarn are regularly rolled upon the pirns from the one end to the other when the machine is in motion. LM are wheels that turn the cylinders which draw the materials forward to the machine when working. ST is a long wooden spout, into which the materials are laid longways, and one end of it introduced at T, betwixt the cylinders U, and is conveyed over the upper part of the wheel K by small cylinders that direct it down between the two rollers M, which convey it into the tubs PR, in the form of a flat narrow string, without any twine or twist.

FIG. 3. Elevation of a Machine by which the prepared materials are drawn into a thread, and at same time gives it a loose or slack twine.

ABCD represent the frame in which the wheels and cylinders revolve. E is a wheel fixed upon the axle G, to drive the wheel F, which is placed upon the perpendicular iron spindle AF. Below the wheel F, upon the same iron spindle, is fastened a conical wheel that turns a wheel which is fixed upon the horizontal iron spindle RS; on this spindle is also fixed two conical wheels, which convey motion to the spindles, pirns, and cylinders. HI is the great cylinder, upon the upper part of which the small cylinders turn; KL is the frame that contains the cylinders. MNOP are the hecks or flyers, which convey the threads of yarn to the pirns. No. 1, 2, 3, 4, are small drum-wheels that conduct the threads to the hecks; and Nos. 5, 6, 7, 8, are the pirns upon which the threads of yarn are rolled by the flyers.

XY is a board that moves the pirns upwards and downwards upon the iron spindles, by means of which the threads are regularly rolled upon the pirns; the flyers that direct the threads to the pirns are fixed upon the upper part of the iron spindles on which the pirns revolve. No. 9, 10, 11, 12, are small weights suspended by straps of leather, which pass over the lower part of the pirns, and regulate their motion, so as to accord with the motion of the flyers.

Fig. 4. Plan, or Bird's-eye View of Fig. 1.

ABCD represents the cast-iron frame; GH are the great cylinders, on the upper part of which the small ones revolve, (see K, Fig. 2.) IK is the spout into which the materials to be spun are laid longways, and one end introduced in betwixt the cylinders (see TU, Fig. 2.) No 1. is a wheel the teeth of which act upon the teeth in the wheel No 2. to turn it round, and No. 2. drives the small wheel No 3. which is fixed upon the iron spindle EF; by this means motion is conveyed to all the wheels and cylinders. UV is a frame, in which the axle of the wheel No. 1. turns when the machine is in motion. LM are two white-iron tubs that receive the materials from the cylinders. NO, PR, and ST, are all tubs made of white-iron for containing the materials both before and after they have past through betwixt the cylinders.

Fig. 5. Plan or Bird's-eye View of Fig. 8.

ABCD represent the cast-iron frame. E is a wheel that drives the wheel F, and a wheel on the same axle with F turns the wheel M, (see 3 and 4, Fig. 1.) which is fixed upon the iron spindle I, and on this spindle is placed wheels that convey motion to all the other wheels, pinions, cylinders, pirns, and flyers, when the machine is in motion (see RS, Fig. 3.) KL are tubs made of white-iron, and from these tubs the drawn-out materials are introduced to the cylinders GH, in order to be drawn out smaller and receive more twine or twist.

PLATE VII.

Fig. 1. Section of a Machine for Spinning Flax, Hemp, &c. or any fibrous material into Yarn, and may be driven either by water, wind, cattle, or by the power of steam, according as it may be found most convenient.

ABCD represent a frame of cast-metal, in which the spinning parts of this machine revolve. cc are the bobbins or pirns upon which the materials to be spun are rolled. EE are iron rollers, by which the roved threads are drawn off these bobbins, passing over the cylinder F and the small rollers GG to the wheels WW, which convey these threads to the pirns HH. By this operation the threads are drawn smaller, and also receive more twine, and are rolled upon the pirns. I I are small weights, suspended by straps of leather, which pass over the lower part of the pirns and regulate their motion; the other ends of these straps are fastened to the board in which the spindles of the pirns HH turn. KK are small rollers fixed upon the iron spindles I, on which the spindles, hecks, and pirns HH is also placed. Z represents a leather belt, coming from a drum-wheel of any moving

PLATE VII.

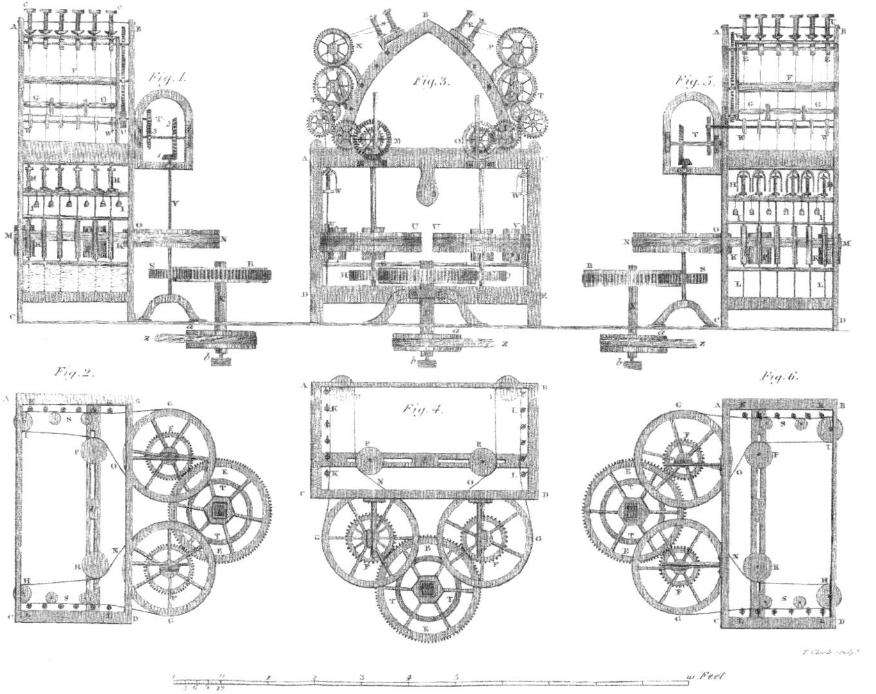

machine, and passing over the drum a, which is fixed upon the axle or shaft X; and upon this axle is also fixed the wheel R, that drives the wheel S, which is fastened upon the iron axle Y; and on this axle is also fixed the drum-wheel O, and over this drum goes the leather belt MN, and turns the rollers KK, with the iron spindles and their pirns HH, upon which the threads of yarn are rolled in the spinning. KP are rollers that revolve upon axles of iron, and these rollers are placed so as to keep the belt MN close upon the rollers KK, to turn them round, with their spindles, hecks, or flyers and pirns. No. 1. is a small pinion fixed upon the axle Y, to turn the wheel No. 2. on the axle of which is fastened No. 3. that drives the wheel T, having fixed upon its axle the pinion U; by this means motion is conveyed from the drum-wheel, a, to the wheels of the cylinders EFGW, and thus they are driven round to convey the materials from the bobbins or pirns at cc down to the pirns HH, upon which the threads of yarn roll in spinning. b is a cod of brass upon which the pivot of the shaft turns when this machine is in motion.

Fig. 2. Plan, or Bird's-eye View of the same machine.

ABCD, represent a frame of cast-metal. EE, is a wheel the teeth of which take upon the teeth of the two wheels FF, to turn them round, and convey motion to the different parts of the machine (see RS, Fig. 1.) G and G, are two drum-wheels fixed upon the axles of the wheels FF; upon one of these drum-wheels goes the belt, GHN, to drive the rollers, LL, which are placed upon the iron spindles with the hecks and pirns, (see HH, Fig. 5.) The belt GOI passing from the other drum, F, turns the rollers KK with their hecks and pirns, on which the threads of yarn are rolled. SS are rollers that revolve upon iron axles, and are placed in the frame so as to hold the belts GHS and GIS close upon the rollers KK and LL. By this means the rollers, with the spindles, hecks, and pirns, are driven round, (see KK, I I, and HH, Fig. 5.) P and R are two rollers which revolve upon axles of iron, placed into longitudinal mortises in the bearer Z; These axles, having liberty to shift in their mortises, can be placed either to tighten or slacken the belts GHN and GOI, according as the pirns require to be driven quick or slow. TT is a drum-wheel, fixed upon the axle of the wheel EE, to receive a belt from a drum-wheel of any machine having power to give it motion. (See Z, Fig. 1. 3. and 5.)

FIG. 3. Cross section of the same machine.

ABCDE represents a frame of cast metal. a is a drum-wheel fixed upon the axle F, and upon this drum is placed the belt Z, that comes from a drum-wheel of any machine in motion. Upon the axle F is also fixed the wheel GG, which drives these two wheels H and I; the two wheels are fixed upon iron spindles, on the upper part of which is placed small pinions to turn the wheels M and O, and thus motion is conveyed to the wheels MTN, and POT, by which the cylinders are turned round when the machine is spinning. RS represent the bobbins upon which the coarse yarn had been previously rolled, and from these bobbins the coarse threads are conveyed by the cylinders down to the pirns WW, to be drawn smaller, and also receive more twine or twist in spinning. UU are two drum-wheels, the one fixed upon the iron spindle of the wheel H, and the other on the spindle of the wheel I; upon these drums are placed the belts which turn the iron spindles with their rollers and pirns, whereon the yarn is rolled in the spinning, (see OMN, and NOM, Figs. 1. and 5.) V and V are the rollers, the axles of which are easily shifted to any side, either to tighten or slacken the leather belts, and cause the iron spindles, with the pirns WW, turn quick or slow as may be found necessary, (see PR, Figs. 2. 4. and 6.) b is a bush, or cod of brass, in which the pivot of the axle F turns when the machine is in motion.

Fig. 4. Plan of the above in Profile, in which ABCD represents the cast-metal frame. EE is a wheel the teeth of which act upon the teeth of the wheels FF to drive them round. GG are two drum-wheels fixed upon the same axles with the wheels FF, (see HU, and UI, Fig. 3.) From one of the wheels, G, the belts HKN pass, to turn the rollers, spindles, and pirns KK; at same time the belt ILO, from the other we turns the rollers, spindles, and pirns LL, (see MNO, and HH, KK, LL, Fig 6.) H is a roller that conducts the belt NH to the small rollers KK, which are fixed upon the iron spindles, and on these spindles is also placed the pirns to receive the yarn when spinning, (see HH, Fig. 6.) I is a roller that conveys the belt OI to the small rollers LL, that are fixed upon the same spindles with the pirns on which the threads of yarn are rolled by the machine in the spinning. TT is a drum-wheel fixed upon the axle of the wheel EE, and over this drum passes a leather belt from a drum of any machine having power and motion to drive this machine.

Fig. 5. Elevation of the same machine. ABCD represent the frame of cast metal. EE are iron rollers which draw the roved materials from off the bobbins or pirns UV, upon which it has been previously rolled, and from these rollers the threads pass over the roller F, the small rollers GG, and through betwixt the wheels WW, down to the hecks and pirns HH, by which the threads receive more twine or twist. IY are weights suspended by leather straps or thongs that pass over the lower part of the pirns HH, and regulate their motion. KK is small roundles fixed upon the iron spindles LL, and on these spindles the hecks or flyers that direct the threads of yarn to the pirns HH are also placed. MN is a leather-belt which passes from the drum-wheel NO over the small roundles KK; by this means the spindles LL, with the hecks and pirns, are turned round. PM is the rollers which direct the belt to the roundles KK. (See IHER, Figs. 2, 4, and 6.)—a, is a drum-wheel fixed upon the axle X; and over this drum-wheel passes the belt Z, which conveys motion to it from a drum-wheel of any moving machine having power to turn it. R is a wheel fixed upon the shaft X, to turn the wheel S, which is fastened upon an iron spindle, and on the upper part of this spindle is also fixed a small pinion that turns the wheel T; thus motion is conveyed to all the other wheels from the drum-wheel, a, to the wheel V; by this means the rollers, EFGW, are turned to convey the threads off from the bobbins VV, down to the pirns HH,—by which operations these threads are drawn smaller, and also receive the twine or twist that may be found necessary.

Fig. 6. Plan, or Bird's-eye View of the same Machine.

ABCD represents the frame that supports the machinery. EE is a wheel which acts upon the two wheels FF, to turn them round, with the drum-wheels GG, which are fixed upon the same axles with the wheels FF; and upon these drum-wheels goes the belts that turn the iron spindles with the flyers, or hecks and pirns. GIO is the belt which turn the spindles with the hecks and pirns KK; and GNH is the belt that turns the spindles with the hecks and pirns LL. (See MN and IH, Fig. 5.) PR are rollers, the axles of which have liberty to shift in long mortises; and by this means the belts can be either tightened or slackened at pleasure, in order to regulate the motion of the spindles. (See PM, Fig. 5.) SS are rollers that hold the belts close upon the roundles of the spindles, to turn them round when the machine is spinning. TT is a drum-wheel fixed upon the axle of the wheel EE, and upon this drum goes a leather belt, passing from the drum-wheel of any machine in motion,—so that this machine can be turned from any engine which is driven either by water, wind, cattle, or steam, according as it may be thought most advantageous.

PLATE VIII.

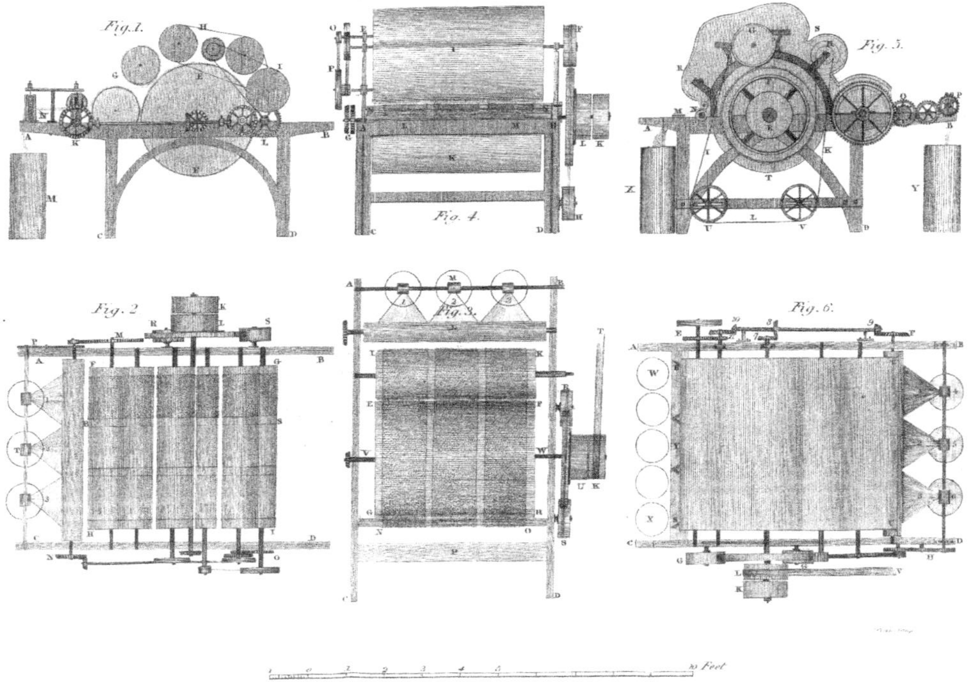

PLATE VIII.

Fig. 1. Elevation of a Machine for Preparing the Refuse of Flax, Hemp, &c. to be spun into coarse Yarn.

ABCD represent a frame of cast-metal, in which the wheels and cylinders revolve. EF is the great cylinder, the circumference of which is covered with leather, and in this leather are fixed bent teeth of small iron-wire, placed near to each other. GHI are lesser cylinders, placed round on the upper part of the great one; the circumference of these cylinders is also covered with leather, in which are likewise fixed teeth of iron-wire; and when the great cylinder is turned round, then the lesser ones are also turned, and by this means the wire-teeth tease or comb the rough materials in passing through betwixt the greater and lesser cylinders. EF and GHI are the lesser cylinders. NKL are wheels by which motion is conveyed from the great cylinder to the lesser ones. (See LM and NO, Fig. 2.) M is one of the tubs made of white-iron, and is placed so as to receive the materials N, which are dropt down at A from the cylinders.

Fig. 2. Plan or Bird's-eye View of the Lesser Cylinders.

ABCD represent the cast-metal frame in which these cylinders revolve. FGRS are parts of the cylinders covered with leather, in which are fixed bent teeth of small iron wire; and RSHI are those parts without any cover. KL are two small drum-wheels placed upon the axle of the great cylinder; the drum, L, being fixed on this axle, over it goes a leather-belt which turns the drum and the great cylinder round, and of course all the other cylinders are put in motion. But, in order that this machine may be easily stopt, the drum-wheel. K, has liberty to revolve upon the axle of the great cylinder, so that when the machine is to be stopt the belt must be shifted off from the drum wheel L, and placed upon the drum-wheel K, which revolves upon its axle and turns with the belt; by this means the drum wheel L, and, of course, the cylinders, are all stopped. MPNO are small wheels which turn the lesser cylinders. T are iron cylinders that draw the combed or carded materials forwards from the cylinders FR. Nos. 1. 2. 3 are triangles made of white iron, having on two sides a moulding or ledge, which bring these fibrous materials into a narrow space, and introduce them in betwixt the iron cylinders T, by which it is drawn forward and dropt into the tubs below. (See AM, Fig. 1.) RS are rollers that conduct the leather-belt to the drum-wheels by which the cylinders are driven. (See UV, Fig. 5.)

Fig. 3. Plan of the Great Cylinder, with four of the lesser ones which revolve in the same horizontal frame.

ABCD represents the cast-metal frame in which the wheels and cylinders turn. EFGH is the great cylinder covered with leather, in which are fixed bent teeth of small iron wire. EFIK is a lesser cylinder, covered with leather, having teeth of wire that convey the materials from the great cylinder at EF to the lesser cylinder L, by which it is laid upon the triangles 1. 2. 3. and from them these materials are introduced betwixt the iron cylinders M, by which they are drawn forwards and dropt into the tubs 1. 2. 3. below. (See NA, Fig. 1.) VW represents the iron axle on which the great cylinder is fixed; and on this axle is also fastened the drum-wheel U, over which goes the leather-belt which drives this cylinder. (See L V, Fig. 6.) K is another drum-wheel placed upon the axle VW, having liberty to revolve on its axle when turned by the belt KT, whilst the drum-wheel U, and of course the great cylinder, is stopt at any time when found necessary. NO are two small cylinders, in the circumference of which are fixed iron-spikes, placed at a small distance from each other, in order to draw the rough materials forward to the cylinders as they revolve. P is a board placed in the iron frame; and over this board the rough materials pass to the small cylinders NO, by which they are introduced to the great cylinder. (See MN, Fig. 6.)

FIG. 4. Profile of the same Machine.

ABCD represents a cast-metal frame, in which the wheels and cylinders are placed. EGOP and FL are wheels fixed upon the axles of the lesser cylinders by which they are driven round when working. LM is a flat board, upon which the rough refuse is spread, and introduced betwixt the small cylinders NN, by which it is conveyed forward to the cylinders to be combed or carded, and thus prepared to be spun into coarse yarn. (See MN, Fig. 6.) IK is a cover of boards which incloses the cylinders. (See RSTO, Fig. 5.) H is one of the rollers that directs the belt to the drum-wheels which turn the cylinders.

FIG. 5. Cross Section of the same Carding Machine.

ABCD represents the frame which carries the wheels and cylinders. E is a drum-wheel fixed upon the axle of the great cylinder, and over it goes a leather-belt which drives this machine. (See L V, Fig. 6.) F is a drum-wheel, which is also fixed upon the axle of the great cylinder; and upon this drum goes the belt IKL, which turns the drum-wheels GH when the drum-wheel F is in motion. X is one of the white-iron tubs into which the rough materials or refuse is placed, and conveyed from this tub over the board M, in between the small cylinders N; by this means the materials are conducted forward to the carding-cylinders; and, after going through betwixt these cylinders, are delivered out again at O, over the cylinders OP, into the tub Y, in the form of a narrow ribbon or garter, without any twine or twist, and are then carried to the spinning-machine, by which this flat thread is drawn smaller, and also receives a slack twine, forming it into a round thread, which is drawn smaller, and likewise receives more twine, from a second machine. RS and TO, is a cover of thin boards, which inclose the carding or combing-cylinders. UV are two drum-wheels that revolve upon axles of iron, placed into long mortises, in which those axles have liberty to shift; and by this means the wheels UV can be placed at a greater or less distance from each other; and thus the leather-belt, IKL, may be either tightened or slackened, according as the lesser cylinders require to be turned quick or slow.

Fig. 6. Plan, or Bird's-eye View of the same Machine.

ABCD represents the frame in which the wheels and cylinders revolve. EF and GH are the wheels and drums which drive the cylinders. No 7. is a conical or bevel wheel, fixed upon the axle of the great cylinder, having teeth to turn the wheel No. 8. which is fastened upon an iron axle; and upon this axle is also fastened the pinion, No. 9. and upon this axle is likewise fixed the pinion No. 10. that drives the wheel No. 11. which is placed upon an axle in one of the lesser cylinders. WX represents white-iron tubs into which the rough materials are placed, and from these tubs are conveyed over the board RYS, in between the cylinders, and placed upon the triangles Nos. 1, 2, 3.; these triangles are made of white-iron, having a ledge fixed on two sides, which contract the fibrous materials into a narrow space, before they reach the iron cylinders BD, which draw them from the cylinders, and conduct them down into the tubs 4, 5, 6, in the form of a narrow flat thread, without any twine or twist. L is a drum-wheel fixed upon the axle of the great cylinders, and upon this drum goes the belt LV, coming from a drum-wheel of any machine having power and motion sufficient to turn the cylinders. But when the cylinders are to be stopped, then the belt LV, must be shifted off from the drum L, and placed upon the drum K, (See KT, fig. 3.) which, having freedom to revolve upon the axle of the great cylinder, is turned round by the belt, whilst the cylinder is stopt from its motion or remains at rest. If the cylinders are to be set a-going, then the leather-belt must be placed upon the drum-wheel L, which is fixed upon the axle of the great cylinder; by this means the lesser cylinders are also driven round.

PLATE IX.

Fig. 1. Plan of a Machine for Spinning the refuse of Flax, &c. into coarse Yarn.

ABCD represent a frame of cast-metal, in which the wheels, flyers, and pirns revolve. E is a wheel fixed upon an iron-axle, (see EP. Fig. 2.) and on the circumference of the wheel, E, are teeth which take upon the teeth of the small wheel F, to drive it round. This wheel is fastened upon an iron spindle, on which is also fixed a pinion to turn a wheel that is placed upon the axle G, by which motion is conveyed to the other wheels. (See No. 1. Fig. 2.) No. 1. 2. 3. are wheels fixed upon the axle G, having teeth to turn the wheels which are placed upon the axle of the drum-wheels, No. 4. 5. 6. HI are the flyers, or hecks, with the pirns upon which the threads of yarn are rolled by the machine when in motion. No. 7. and 8. are a wheel and endless screw, by which motion is conveyed to the frame that moves the pirns up and down, to cause the threads of yarn roll regularly upon them. KL and MN are tubs made of white-iron, into which is placed the materials that are to be spun into yarn.

PLATE IX.

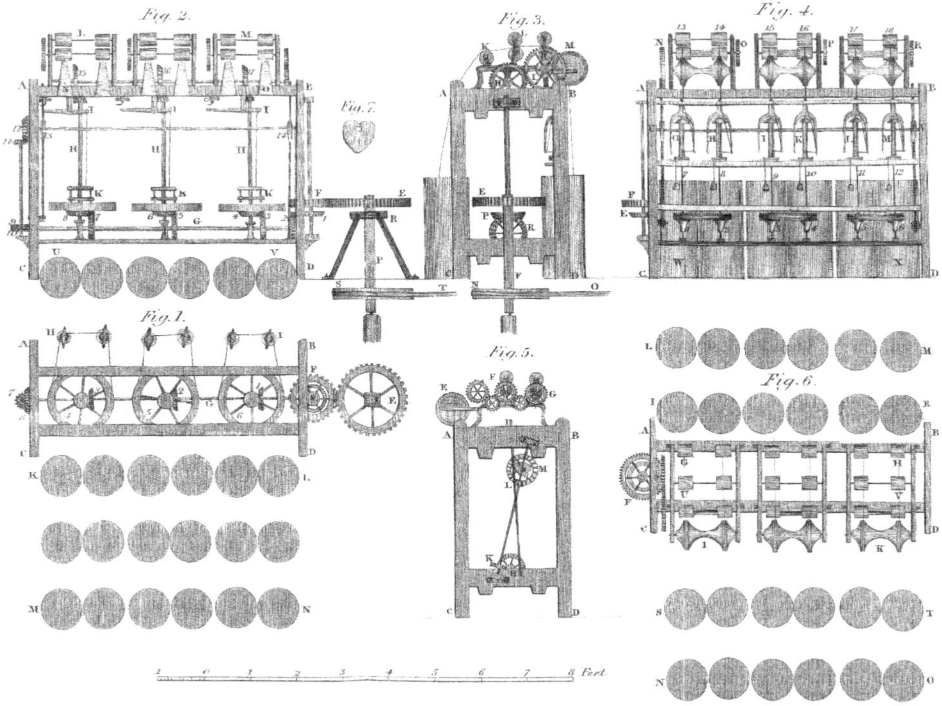

FIG. 2. Profile of the same Machine.

ABCD is the cast-metal frame that carries the wheels, flyers, and pirns. E is a wheel fixed upon the iron-axle P, and on this axle is also fastened the drum-wheel S, upon which is placed the leather-belt T, to turn this wheel, by the belt extending from a drum-wheel of any machine in motion which is driven by water, by wind, cattle, or steam. Upon the circumference of the wheel E are teeth that act upon the teeth of the wheel F to drive it round. This wheel is fixed upon the perpendicular iron spindle BF; and upon this spindle is also fastened the wheel No. 1. that turns the wheel No. 2. which is fixed upon the horizontal iron axle G; and on this axle is likewise fastened the wheels No. 3. 5. 7. to drive the wheels No. 4. 6. 8. which are fixed upon the perpendicular iron-spindles HHH; and on the upper part of each spindle is placed a small pinion having teeth to turn the wheels No. 15. 16. 17. and convey motion to the cylinders LM, by which the threads of yarn are drawn smaller, and directed to the flyers, or hecks, to receive more twine, or twist, and also rolled upon the pirns. I I I are levers, each of which turns upon an iron bolt placed in the frames; to the shorter arm of each lever is attached an iron rod, as IK, IK, IK, so that by pushing down the longer arm of these levers the cranks, K, K, K, is raised clear of the drum-wheels 4. 6. 8.; by this means any two of the flyers wit the pirns can be stopt at pleasure. No. 9. is an endless screw fixed upon the horizontal iron axle G, to turn the wheel No. 10. which is fastened upon the lower end of an iron spindle; and upon its upper part is fixed the pinion No. 11. to turn the wheel No. 12. which is fixed upon one end of a horizontal iron axle, and on this axle is also fastened two eccentrical wheels, or hearts, one at No. 13. and another at 14. (see Fig. 7.) and when the iron axle with these wheels are turned round, then the wheels move the upper ends of two perpendicular iron rods forward and backward, their lower ends having liberty to turn on iron bolts in the frame, (see I LH, Fig. 5.) By this means the board, upon which the lower end of the pirns turn, is raised up and down at every revolution of these eccentrical wheels; thus the threads of yarn are rolled regularly upon the pirns, from one end to the other, when the machine is in motion. R is a coupling brass placed upon the top of an iron frame, and in this brass the upper journal of the axle P revolves, whilst the pivot in the lower end turns in a cod of brass. UV represent the tubs of white-iron into which the materials to be spun are placed.

Fig. 3. Cross section of the same machine.

ABCD represents the frame in which the wheels, flyers, and pirns revolve. E is a wheel fixed upon the iron axle F, and upon this axle is also fastened the drum-wheel N, over which goes the leather belt NO, that turns this wheel round. C is one of the white-iron tubs into which the materials to be spun are placed, and from this tub it goes over a roller at A, to the cylinders KLM, by which the threads are conveyed to the flyers and pirns at B, to receive more twine or twist, and rolled upon the pirns. HI are wheels which turn the cylinders. P is a conical wheel, fixed upon an iron axle, to turn the wheel R. (See Nos. 1. and 2. Fig. 2.) D is a white-iron tub into which the pirns are put when full of yarn, and in which tub it is conveyed to the reeling machine.

FIG. 4. Elevation of the same machine.

ABCD represent the cast-metal frame in which the wheels, cylinders, flyers, or hecks and pirns, revolve. EF are the wheels by which motion is carried to the other wheels. (See F and No 1. Fig. 2.) Nos. 1, 2, 3, 4, 5, 6, are round sheaves fixed upon the perpendicular iron spindles 7, 8, 9, 10, 11, and 12. Upon the top of these spindles are fastened the flyers or hecks, GHIKLM, which direct the threads of yarn upon the pirns, and on these spindles the pirns revolve, whilst the spindles are turned round by a leather belt from the drum-wheels Nos. 2, 4, 6. (See HI, and 4, 5, 6. Fig. 1.) NOPR represents the wheels that turn the cylinders 13, 14, 15, 16, 17, and 18, which convey the threads down to the flyers. UV is an iron axle, and upon it is fixed two eccentrical wheels, the one at U and the other at V, so that when this axle is turned round, these wheels move the board up and down upon which the lower end of the pirns turn; by this means the threads of yarn roll regularly upon the pirns, from the one end to the other. The balls Nos. 7, 8, 9, 10, 11, and 12. represent small weights suspended by leather straps, the other end of which are fastened to the board whereon the pirns turn. These leather straps go over the lower round of the pirns, and regulate their motion, according as the weights are hung at a greater or less distance from the centre of the pirns. WX is the tubs into which the materials are placed, and conveyed by the cylinders to the flyers, which direct it to the pirns, whereon it is rolled when the machine is in motion.

Fig. 5. Cross section of the same machine.

ABCD represents the cast-metal frame that carry the wheels and cylinders, with the flyers and pirns. EFG are wheels that turn the cylinders which draw the materials forward to be spun. (See KLMB, Fig. 3.) IH is a perpendicular rod of iron, the lower part of which turns on an iron bolt at H, and its upper end, I, is moved forward by the eccentrical wheel, or heart L, and backward by a weight suspended at a cord, the other end of which is fixed to the iron rod at I.

Thus the board on which the pirns turn is moved up and down to cause the threads of yarn roll regularly upon the pirns. HK is the wheel that conveys motion to the wheels, cylinders, flyers, and pirns. (See 1. 2. 3. Figs. 1. and 2.)

FIG. 6. Plan or Bird's-eye View of the same Machine.

ABCD is the cast-metal frame; F the wheels that convey motion to the cylinders, flyers, and pirns. LMPR are tubs made of white-iron; in these tubs the materials are placed, and from them introduced to the cylinders GHUV that draw the coarse threads forward to the cylinder IK, which conducts them to the flyers and pirns whereon they are rolled by the machine. (See GHIKLM, Fig 4.) NOST are the tubs which receive the pirns when full of yarn, and in these tubs they are carried to the reeling-machine.

PLATE X.

Fig. 1. Profile of the Reeling Machine.

ABC represent a frame of cast-metal. D is a round wooden axle, into which is fixed the spokes or arms 1, 2, 3, 4, 5, 6; and upon the extremity of these arms are fastened longitudinal spars of wood. (See IKLM, Fig. 2.) E is the pirn placed upon an iron spindle fixed in the frame upon which the pirns revolve, whilst the thread, I, is rolling round on the extremity of the reel.

Fig. 2. ABCD represents a cast-metal frame in which the reel revolves. EF is an axle of wood, wherein is fixed the arms or spokes, upon the extremity of which are fixed the longitudinal spars IKLM; and upon the outer side of these spars the threads of yarn are rolled. NOP represent three of the arms, having liberty to shift inwards or outwards in their mortises; so that when the hanks of yarn are completed, then the arms, NOP, are shifted inwards, by which means the hanks of yarn are slackened, and of course easily taken off the reel. But when the

PLATE X.

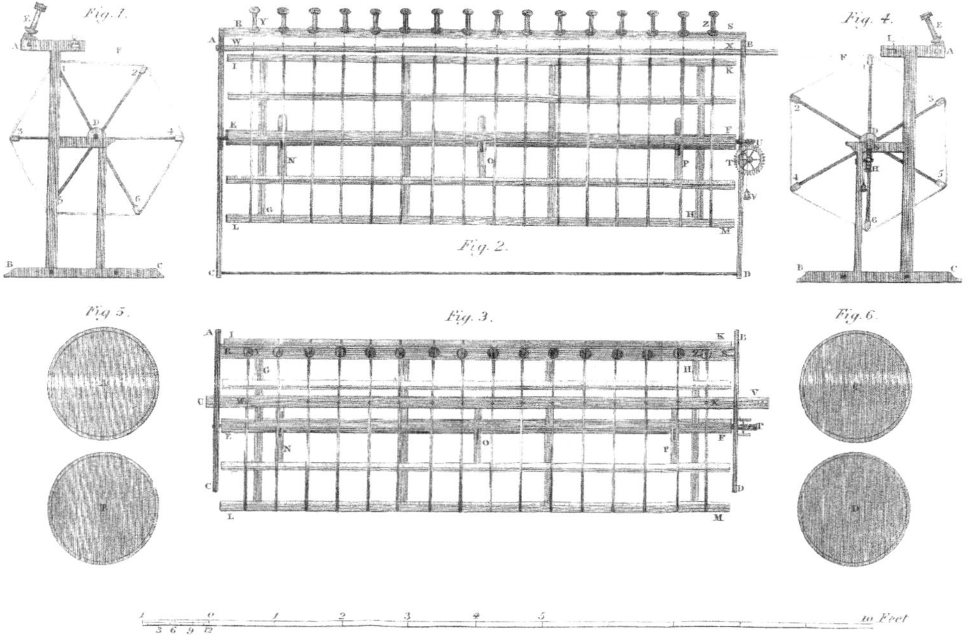

T. Clerk Sculp.

reeling is to begin, these arms must be shifted outwards again, so that the spar which is fixed on the outer ends may be at the same distance from the centre of the axle, EF, as that of the other five spars of the reel. In each of these shifting arms is a mortise to receive a small wedge that rests on the outside of the axle, and prevents them from slipping inward in the time of reeling; but when the hanks of yarn are to be taken off, these wedges must be taken out. RS is a wooden rail, and in it is fixed round iron spikes, upon which the pirns YZ revolve when the reeling is going on. U is an endless screw upon the pivot of the axle F. This screw takes into the teeth of the wheel T to turn it round; and in this wheel are one hundred and twenty teeth, so that every revolution of the reel moves the wheel T one teeth forward, and one turn of this wheel makes one hundred and twenty threads round the reel; at this instant the bell V being raised a little by the wheel, T, gives a tinkle of alarm, to shew that the reel has performed a hundred and twenty revolutions, and that the cut is complete. This number of threads makes one cut; and twelve of these cuts make what is termed a hank or hesp; and four of these hanks make a spindle. Thus, one hundred and twenty threads round the reel makes one cut, each thread measuring seven feet six inches, or two and a half English yards; so that one cut of yarn measures 900 feet in length; one hesp or hank 10,800 feet; and one spindle, or four hanks, measures 43,200 English feet, or 14,400 yards in length. WX is a spar of wood, in which is fixed, opposite to each pirn, a small iron bolt, having a hole in its upper part; through these holes the threads pass from the pirns, and by this means are directed to the reel. When one cut is complete, it is tied with a thread, and the spar, WX, is shifted a little on end, to prevent the threads from rolling on above the other cut; thus the spar WX must be shifted a little on end when each cut is reeled, till the twelve cuts are complete, and then the hank is taken off the reel.

FIG. 3. Plan, or Bird's-eye View of the same Machine.

ABCD is the frame in which the reel revolves. EF is the axle into which the spokes or arms of the reel are fixed. GH are the arms of the reel, and IKLM the spars fixed upon the extremity of these arms. NOP is the shifting arms, which have liberty to slide inward in their mortises, to slacken the hank of yarn, so that it may be easily taken off the reel. RS is a wooden rail, and into it is fixed round iron spikes, upon which the pirns YZ revolve, when the machine is in motion. WX is a rail, in which are fixed small iron bolts; and in the top of these bolts is an opening through which the threads of yarn pass to the reel. T is an endless screw on the pivot of the axle F. This screw turns the small wheel which counts or numbers the threads of yarn in each cut when reeling. The rail WX has liberty to shift from V towards U; and must be shifted a little when each cut is reeled, in order to prevent the threads from rolling above the one which was reeled before,—so that the cuts may be separate from one another upon the reel, and also a little distance between the hanks when complete, in order that they may be readily taken off the reel.

FIG. 4. Cross Section of the same Machine.

ABC represents a cast-metal frame upon which the pivot, G, of the reel turns. (See FU, Fig. 2.) D is an axle, into which is fixed the arms or spokes, 1, 2, 3, 4, 5, 6; and upon the extremity of these arms are fastened longitudinal spars of wood, upon the outside of which the thread, I, of the yarn is rolled, as it comes off the pirn, E, (See WL, Fig. 2.) In the frame at A, is fixed round iron spikes upon which the pirns revolve, (See YZ, Fig. 2.) F represents the yarn rolled round the circumference of the reel. G is an endless screw, upon the pivot of the axle D of the reel, the spirals of which take in between the teeth of the wheel H, to turn it round, and number the threads of yarn in each cut when reeling.

Figs. 5. and 6. are baskets made of willow, or small twigs, in which the pirns, when full of yarn, are carried from the spinning machine to the reel; and in these baskets the empty pirns are conveyed from the reel to the spinning machine, alternately, whilst the spinning and reeling machines are employed.

POSTSCRIPT.

LINEN MANUFACTURE OF IRELAND.*

THE foundation of the linen manufacture in Ireland was laid by the unfortunate Earl of Strafford, during the time he resided in that country as chief governor.† Having observed

* See WAKEFIELD's Account of Ireland, Vol. I. p. 680.

† The following is an extract from his Lordship's letter on this subject, giving an account of the report he had made to the king and council. The letter is dated July 25, 1636.

"That there was little or no manufactures among them, but some small beginnings towards a clothing trade, which I had, and so should still, discourage all I could, unless otherwise directed by his Majesty and their lordships; in regard it would trench not only on the clothings of England, being our staple commodity, so as if they should manufacture their own wool, which grew to very great quantities, we should not only lose the profit we made now by in-dressing their wools, but his Majesty lose extremely by his customs, and, in conclusion, it might be feared they might beat us out of the trade itself, by underselling, in which they were able to do. Yet I have endeavoured another way to set them on work, and that is, by bringing in the making of linen cloth; the rather in regard to the women, who are naturally bred to spinning; that the Ir-

that the soil, in many parts, was suited to the production of flax, that the women were chiefly bred to spinning, and that the price of labour was cheap, he conceived that linen might be made there at such an expense as would enable the manufacturers to undersell, twenty per cent at least, those of Holland and France. Impressed with this idea, he sent to Holland for flax-seed, and to the Netherlands and France for competent workmen. The flax was sown, and succeeded according to expectation; spinners and looms were set to work; and his lordship, to animate others, embarked himself in the business, and expended in promoting it, L.30,000 of his private fortune *.

By the troubles, however, which soon after broke out, an entire stop was put to this useful enterprise, until it was again revived by the Duke of Ormonde, who procured several acts of par-

ish earth is apt to the bearing of flax, and that this manufacture would be, in conclusion, rather a benefit than other to this kingdom. I have, therefore, sent for the flax-seed into Holland, being of a better sort than we have, and have sown this year a thousand pounds of it, (finding, by some I sowed the last year, that it takes very well here.) I have sent for workmen out of the Low Countries, and south of France, and set up already six or seven looms, which, if it pleases God so to bless us this year, I trust so to invite them to follow it, when they see the great profit arising thereby, as that they shall generally take and employ themselves that way, which, if they do, I am confident it will prove a mighty business."

* Carte's Life of the Duke of Ormonde, vol. ii. p. 85.

liament to encourage it, and also for inviting protestant strangers to settle in the kingdom. He dispatched skilful persons to the Netherlands, to make themselves acquainted with the method of cultivating flax in that country; the manner of preparing and weaving it, as well as the process employed in bleaching the cloth, and likewise to engage workmen experienced in the different branches of the art to settle in Ireland. At the same time he wrote to Sir William Temple, whom he had recommended to the king, as a man fit for the purpose, and who was then ambassador at Brussels, to send him over from Brabant 500 families, who had been employed in this manufacture: he procured others from Rochelle, and the Isle of Rhè; and Sir George Carteret supplied him with a considerable number from Jersey, and the adjacent parts of France. He built tenements for some of these people at Chapel Izod, near Dublin, where, in a little time, 300 hands were at work, under the direction of Colonel Lawrence. Cordage, sail-cloth, ticking, and linen, made of Irish yarn, equal in quality to any in Europe, and which could be sold at a less price than that of any other country.

His Grace erected another manufactory at his own town of Carrick, assigning to the workmen one half of the houses in the place, with 500 acres of land contiguous to the walls, for three lives, or thirty-one years, at a pepper-corn for the first, and afterwards at two-thirds of the old rent. All plans of improvement among a people just beginning to emerge from barbarity, meet with formidable enemies from prejudice and inveterate habits. It was not therefore to be expected, that this new branch of industry could be introduced among the Irish without considerable labour and exertion. The indolence of the inhabitants, and their ignorance of the art, were the greatest obstacles to the progress of the infant manufacture; but the Duke's perseverance surmounted every difficulty; by his example and encouragement, a spirit of enterprise, to which the country had before been a stranger, was at length excited, and he had the satisfaction, before he quitted the government in the year 1669, of seeing the linen business fully established, and in a promising condition.*

The importance of this branch of industry, soon attracted the attention of parliament, and considerable sums of money were from time to time voted for its support. That the business might be better superintended, a board of trustees, who first met in 1711, was established †, and in order that no encouragement might be wanting, bounties on the exportation of Irish

* Carte's Life of the Duke of Ormonde, vol. ii. p. 342.

† Crump's Essay on the best Means of providing Employment for the Poor, p. 75.

linen were offered, and in 1743 they were distributed. *

Under the direction of the above-mentioned board, the cambric manufactory was also introduced in 1737, by a Mr De Joncourt, who brought over workmen for that purpose from France. It was first established at Dundalk, on the estate of Lord Viscount Limerick, afterwards Earl of Clanbrassil, and being supported by a voluntary contribution of L.30,000, was, in the year 1770, in a very thriving condition. † The author of a pamphlet on the Absentees of Ireland, published at Dublin in 1767, says, that at that period, there were upwards of one thousand five hundred looms employed in weaving cambric in the northern part of the country. ‡

The preparation of the flax plant, and the various operations it undergoes before it is made into cloth, gives employment to so many people in Ireland, and brings into the national coffers so large an annual income, that the linen which is formed from it has, with great propriety, been called the staple manufacture of the country. To give the reader a clearer idea of its extent and importance, I shall subjoin the following table,

* Lord Sheffield's Observations on the Manufactures of Ireland, p. 75.

† Boulter's Letters, Vol. II. p. 166.

‡ List of the Absentees in Ireland, with Observations on the Trade, Manufactures, &c. p. 67.

which shews the number of acres sown with flax in the year 1810. It is copied from a return made to the trustees of the linen board; but it appears, by a note of the secretary, that it includes pieces of land less than an acre, and as the quantity of small patches employed in raising flax in Ireland must be immense, I am of opinion, that they may be fairly estimated at least at 20,000 acres. If this number be added to the result given by the table, the sum total will amount to about 100,000 acres, which allowing the average produce to be 30 stone per acre, at 10s. 6d. per stone, the average price in the last seven years will give an annual produce of the raw material worth L. 1,500,000.

	COUNTIES.	Total Number of Acres supposed to be sown with Flax in the Year 1809.	Total Number of Bushels of Flax-Seed supposed to have been saved out of the Crops, without distinction whether the whole of the Bounty will be claimed thereon or not.	Estimated Portion of the Total Number of Bushels contained in the preceding Column, on which Bounty is likely to be claimed.	Rate of Bounty per Bushel.	Amount.		
		Acres.	*Bushels.*	*Bushels.*		£	*s.*	*d.*
Ulster.	Antrim	11,100	3,100	2,900	5*s.*	725	0	0
	Armagh . . .	15,000	6,000	6,000	5*s.*	1,500	0	0
	Londonderry . .	5,000	3,200	3,000	5*s.*	750	0	0
	Tyrone	10,241	8,460	7,000	5*s.*	1,750	0	0
	Donegal . . .	6,000	7.200	5,600	5*s.*	1,400	0	0
	Fermanagh . .	5,000	5,000	4,400	5*s.*	1,100	0	0
	Monaghan . . .	3,200	4,000	3,600	5*s.*	900	0	0
	Cavan	4,300	6,500	700	5*s.*	175	0	0
	Down	2,700	3,200	3,000	5*s.*	750	0	0
Leinster.	Meath	112	842	840	5*s.*	210	0	0
	Louth	1,253	11,590	10,431	5*s.*	2,607	0	0
	King's County .	252	2,000	1,000	5*s.*	250	0	0
	Longford . . .	1,325	250	240	5*s.*	60	0	0
	Kildare and Wicklow	36	100	80	5*s.*	20	0	0
	Queen's County .	18	114	48	5*s.*	12	0	0
	Kilkenny . . .	50	350	50	5*s.*	12	10	0
	Westmeath . .	950	3,000	800	5*s.*	200	0	0
	Carlow and Wexford	111	500	280	5*s.*	70	0	0
	Dublin	- -	- -	- -	- -	-	-	-
Munster.	Cork	1,462	4,481	1,794	5*s.*	448	10	0
	Clare	58	340	340	5*s.*	85	0	0
	Limerick . . .	1,500	1,100	400	5*s.*	100	0	0
	Kerry	554	1,572	1,572	5*s.*	393	0	0
	Tipperary . . .	117	1,600	822	5*s.*	205	0	0
	Waterford . . .	25	63	19	5*s.*	4	15	0
Connaught.	Sligo	683	1,800	600	5*s.*	150	0	0
	Mayo	1,338	5,308	4,800	5*s.*	1,202	0	0
	Galway	1,196	9,568	8,372	5*s.*	2,093	0	0
	Leitrim	1,565	1,300	1,000	5*s.*	250	0	0
	Roscommon . .	1,707	1,707	1,707	5*s.*	426	15	0
	Total . .	76,749	94,145	71,405		17,851	5	0

A very intelligent manufacturer,* who was applied to for information on this subject, in a memorandum transmitted to one of my friends, says, " The spinning of linen yarn by machinery was not practised in Ireland till within these few years; this improvement was introduced into the county of Down, by Mr Cruikshank, who first made the experiment on a very contracted scale. He afterwards tried it at Cork, and at present, (May, 1811,) there are several manufactories on this principle in the northern part of the country, though not so many as might have been expected.

" The leading cause against the extension of machinery, is the low price of labour; yarn spun by women is sold here much cheaper than the same article manufactured by machinery in England. In the latter country, the labouring classes have a choice of employments, in consequence of the variety of its manufactures. If one fail them, they can apply to another; with us the case is different; when trade is bad, its worst effects are felt by the poor women, who must sell at the prices of the day, or remain unemployed; and it frequently happens, that they dispose of the worked article for less than the raw materials cost them.

* Mr Joseph Nicholson, of the county of Armagh.

" To one unacquainted with Ireland, the small earnings of the poorer females, frequently not more than two-pence per day, working diligently from morning till night, for months together, must appear very extraordinary, and under such circumstances, it is unlikely that this trade should increase so much as it might, though spurred on in the beginning by offers of large premiums from the linen board.

" With the aid of machinery, one person is fully equal to do the work of ten, and in this the advantage in its favour appears to be great ; yet when it is considered that women are fully competent to bring any description of yarn which we spin, to double the fineness that machinery can, this advantage is much lessened.

" Another thing against us is, that we cannot with any material, let its quantity be what it may, exceed the fineness of three hanks, or thirty-six leas in the pound ; whereas women, when the flax is good, can run it from twelve to twenty hanks, or from 144 to 240 leas in the pound.

" Our yarn, from the better quality of the raw material is superior ; it sells, therefore, considerably higher, and is commonly used as warp for yarn spun by the hand.

"On balancing the account, we believe the advantages are in favour of mill-spinning; and we are of opinion, that this improvement will one time or other become a source of prosperity to the country.

"The wages in the factories may on an average be eight-pence per day, and spinners could not be procured for much less; in their own houses they are satisfied with four pence or five-pence.

"To sum up the whole, the quantity of yarn spun by machinery, the greater price given for it, and the better means of obtaining a market, are in favour of that mode with respect to coarse yarn. On the other hand, the low price of labour, the superior fineness of the wrought material, and the expence of machinery, with its wear and tear, are very great drawbacks; yet it may be estimated, that the balance is in favour of the former, to the amount, it is supposed, of one halfpenny per hank, or twelve leas.

"The usual earning of weavers in this country is from six shillings to eight shillings per week."

Spinning by the hand, in a country like Ireland, where the sedentary occupations of the female must be so often interrupted, not only by domestic, but agricultural labours, is attended with one very striking advantage, namely, that it can be suspended and resumed without any inconvenience, and thus small intervals of time are filled up that would otherwise be lost. A very sagacious writer, says, ' There can be no doubt, but all kinds of employment that can be followed without prejudice from interruption; work that can be taken up and laid down often in a day without damage, such as spinning, knitting, weaving, &c. are highly advantageous to a community, because in them may be collected all the produce of those fragments of time that occur in family business, between the constant and necessary parts of it that usually occupy females; as the time between rising and preparing for breakfast, between breakfast and preparing for dinner, *&c.* The amount of all these fragments is, in the course of a year, very considerable to a single family; to a state proportionably highly profitable; therefore it is, in this case, also to follow that divine direction, *gather up the fragments that nothing may be lost.* Lost time is lost subsistence; it is therefore lost treasure.

" Hereby, in several families many yards of linen have been produced from the employment of those fragments only, in one year, though such families were just the same in number as when so employed." *

* Franklin's Polit. Fragments in his Works.

To this it may be added, that the women in Ireland seem to have a peculiar talent for this kind of work. At least, such is the opinion of a very cèlebrated writer, who observes, that "of all women the Irish are the aptest, and best calculated to spin linen thread well, who labouring little in any kind with their hands, have their fingers more supple and soft than others of as poor condition amongst us." *

Yarn spun by the hand is carried to different degrees of fineness, in different parts of the country, according to the nature of the manufactories which have been established there, and the quality of the flax which they produce; but in this respect, female labour has certainly the advantage over machinery, as the yarn produced by the former may attain to a degree of fineness which cannot be given to it by the latter. Some instances of the ingenuity of the Irish women in spinning are extraordinary; and the following account, extracted from the Belfast magazine, of the performances of Anne M'Quillin, of Comber, in the barony of Castlereagh, county of Down, which seems, in some measure, to confirm the idea of Sir William, deserves to be recorded †. "Perhaps it may not be amiss to mention, that

* Temple's Miscellanies, edit. 1681. p. 13.

† For March 1809.

Anne cannot spin her superfine yarn except in mild weather. Frost, high winds, and excessive drought, are equally unfavourable; she then spins what she calls coarse yarn, that is from 24 to 30 or 36 hanks in the pound. A few years ago, she could spin a hank of 64 in the pound in four days. Of the finer staple she can now spin a hank in the week when the weather is favourable; but the superfine, that is, about 100 hanks in the pound, requires nearly twice the time. As what I have written may reach the eye of some incapable of estimating the fineness of Anne's yarn, from their ignorance of its count and measure, I shall add, that each cut contains 120 threads, and each thread measures 2½ yards. Hence it is evident, that the thread of her yarn, of different hanks from the pound of flax, is in length,

	Miles.	*F.*	*P.*	*Yds.*	
64 hanks,	102	6	34	2	Irish measure.
	130	7	7	1	English ditto.
85 hanks,	136	4	34	2	Irish.
	173	6	36	2	English.
and 105 ditto per lb.	168	6	0	0	Irish.
no less than	214	6	7	0	English *.

* Let the reader compare this with the following account of German spinning: "The peasants in the neighbourhood of

" To the female nobility and gentry of Ireland, Lady Dufferin has set a noble and praiseworthy example; she has repeatedly purchased yarn, had it manufactured into thread by Anne M'Quillin herself, and wrought into lace in the neighbourhood of London. I have in my possession, at this moment, a specimen of the thread, manufactured from yarn, of about 70 hanks in the pound; and lace work from her ladyship from it, is far superior to any thing of the kind which was ever seen in this country. Would the other ladies of the

Bielfield, Gutterslohe, and Rittberg, in Westphalia, who wear wooden shoes, spin, with their clumsy fingers which hold the plough, yarn of such a fineness, that a pound of flax produces a thread twenty-three miles in length, each mile of 20,000 feet, (about 88 English miles,) and that sixteen cuts of yarn, or 19,200 threads, each six feet in length, can, when pressed together, be drawn through a common finger-ring." *Archiv for Statistik Politik og Huusholdnings Videnskaber*, 4 hefte, p. 333.

However strange it may appear to a man of the lower class in Ireland, it is a well-known fact, that the male part of the peasantry, in some districts of Germany, employ their leisure time in spinning. Sussmilch says: " I always reflect with pleasure on the laborious Silesians, and the inhabitants of the neighbouring provinces. These people are temperate in their living, and industrious; in winter they do not sleep away neir time, but all spin, men as well as women, without any distinction of sex; none of them are ashamed of this occupation. The peasants in the Mark are of a different way of thinking, and much less industrious; the men would consider spinning as a disgrace, and spend the dark evening and morning hours either in bed or hanging over a warm stove." *Die Gottliche Ordnung in den Veranderungen des Menschlichen Geselechts*, Berlin, 1775, vol. ii. p. 37.

land follow her steps in the management of our Irish spinstresses, they would call forth Irish ingenuity, render Irish industry active and comfortable, and, by the splendour of their virtues, add real dignity to their titles. *Nobilitas sola est atque unica Virtus.* W. STEELE DICKSON."

There are many parts of Ireland where the manufacture extends no farther than to spinning, markets being held in these places for yarn only. The quality of the linen depends upon the fineness of the yarn, and on that account the finest is manufactured in those parts where the finest yarn is spun. In many instances, the flax is raised, spun into yarn, and woven into cloth by the same person and his family. Taking Ireland, however, in general, the cultivators of flax are much more numerous than the spinners, as is shewn by the exportation of unspun flax; and the spinners out-number the weavers, as appears by the same list, in the exportation of linen yarn. Linen is woven of different widths, from ten-inch bandle linen made in Kerry to 5-4 sheetings manufactured in the neighbourhood of Coote-hill, and of different qualities, from coarse thin 3-4 wides in Antrim, which sell at sixpence per yard, to cambrics worth one guinea.

THE END.

Made in the USA
Monee, IL
04 November 2026

67385008R00075